Condensed Matter Physics: A Very Short Introduction

VERY SHORT INTRODUCTIONS are for anyone wanting a stimulating and accessible way into a new subject. They are written by experts, and have been translated into more than 45 different languages.

The series began in 1995, and now covers a wide variety of topics in every discipline. The VSI library currently contains over 700 volumes—a Very Short Introduction to everything from Psychology and Philosophy of Science to American History and Relativity—and continues to grow in every subject area.

Very Short Introductions available now:

ABOLITIONISM Richard S. Newman
THE ABRAHAMIC RELIGIONS Charles L. Cohen
ACCOUNTING Christopher Nobes
ADDICTION Keith Humphreys
ADOLESCENCE Peter K. Smith
THEODOR W. ADORNO Andrew Bowie
ADVERTISING Winston Fletcher
AERIAL WARFARE Frank Ledwidge
AESTHETICS Bence Nanay
AFRICAN AMERICAN RELIGION Eddie S. Glaude Jr
AFRICAN HISTORY John Parker and Richard Rathbone
AFRICAN POLITICS Ian Taylor
AFRICAN RELIGIONS Jacob K. Olupona
AGEING Nancy A. Pachana
AGNOSTICISM Robin Le Poidevin
AGRICULTURE Paul Brassley and Richard Soffe
ALEXANDER THE GREAT Hugh Bowden
ALGEBRA Peter M. Higgins
AMERICAN BUSINESS HISTORY Walter A. Friedman
AMERICAN CULTURAL HISTORY Eric Avila
AMERICAN FOREIGN RELATIONS Andrew Preston
AMERICAN HISTORY Paul S. Boyer
AMERICAN IMMIGRATION David A. Gerber
AMERICAN INTELLECTUAL HISTORY Jennifer Ratner-Rosenhagen
THE AMERICAN JUDICIAL SYSTEM Charles L. Zelden
AMERICAN LEGAL HISTORY G. Edward White
AMERICAN MILITARY HISTORY Joseph T. Glatthaar
AMERICAN NAVAL HISTORY Craig L. Symonds
AMERICAN POETRY David Caplan
AMERICAN POLITICAL HISTORY Donald Critchlow
AMERICAN POLITICAL PARTIES AND ELECTIONS L. Sandy Maisel
AMERICAN POLITICS Richard M. Valelly
THE AMERICAN PRESIDENCY Charles O. Jones
THE AMERICAN REVOLUTION Robert J. Allison
AMERICAN SLAVERY Heather Andrea Williams
THE AMERICAN SOUTH Charles Reagan Wilson
THE AMERICAN WEST Stephen Aron
AMERICAN WOMEN'S HISTORY Susan Ware
AMPHIBIANS T. S. Kemp
ANAESTHESIA Aidan O'Donnell
ANALYTIC PHILOSOPHY Michael Beaney
ANARCHISM Alex Prichard

ANCIENT ASSYRIA Karen Radner
ANCIENT EGYPT Ian Shaw
ANCIENT EGYPTIAN ART AND ARCHITECTURE Christina Riggs
ANCIENT GREECE Paul Cartledge
THE ANCIENT NEAR EAST Amanda H. Podany
ANCIENT PHILOSOPHY Julia Annas
ANCIENT WARFARE Harry Sidebottom
ANGELS David Albert Jones
ANGLICANISM Mark Chapman
THE ANGLO-SAXON AGE John Blair
ANIMAL BEHAVIOUR Tristram D. Wyatt
THE ANIMAL KINGDOM Peter Holland
ANIMAL RIGHTS David DeGrazia
ANSELM Thomas Williams
THE ANTARCTIC Klaus Dodds
ANTHROPOCENE Erle C. Ellis
ANTISEMITISM Steven Beller
ANXIETY Daniel Freeman and Jason Freeman
THE APOCRYPHAL GOSPELS Paul Foster
APPLIED MATHEMATICS Alain Goriely
THOMAS AQUINAS Fergus Kerr
ARBITRATION Thomas Schultz and Thomas Grant
ARCHAEOLOGY Paul Bahn
ARCHITECTURE Andrew Ballantyne
THE ARCTIC Klaus Dodds and Jamie Woodward
HANNAH ARENDT Dana Villa
ARISTOCRACY William Doyle
ARISTOTLE Jonathan Barnes
ART HISTORY Dana Arnold
ART THEORY Cynthia Freeland
ARTIFICIAL INTELLIGENCE Margaret A. Boden
ASIAN AMERICAN HISTORY Madeline Y. Hsu
ASTROBIOLOGY David C. Catling
ASTROPHYSICS James Binney
ATHEISM Julian Baggini
THE ATMOSPHERE Paul I. Palmer
AUGUSTINE Henry Chadwick
JANE AUSTEN Tom Keymer
AUSTRALIA Kenneth Morgan
AUTISM Uta Frith
AUTOBIOGRAPHY Laura Marcus
THE AVANT GARDE David Cottington
THE AZTECS Davíd Carrasco
BABYLONIA Trevor Bryce
BACTERIA Sebastian G. B. Amyes
BANKING John Goddard and John O. S. Wilson
BARTHES Jonathan Culler
THE BEATS David Sterritt
BEAUTY Roger Scruton
LUDWIG VAN BEETHOVEN Mark Evan Bonds
BEHAVIOURAL ECONOMICS Michelle Baddeley
BESTSELLERS John Sutherland
THE BIBLE John Riches
BIBLICAL ARCHAEOLOGY Eric H. Cline
BIG DATA Dawn E. Holmes
BIOCHEMISTRY Mark Lorch
BIOGEOGRAPHY Mark V. Lomolino
BIOGRAPHY Hermione Lee
BIOMETRICS Michael Fairhurst
ELIZABETH BISHOP Jonathan F. S. Post
BLACK HOLES Katherine Blundell
BLASPHEMY Yvonne Sherwood
BLOOD Chris Cooper
THE BLUES Elijah Wald
THE BODY Chris Shilling
NIELS BOHR J. L. Heilbron
THE BOOK OF COMMON PRAYER Brian Cummings
THE BOOK OF MORMON Terryl Givens
BORDERS Alexander C. Diener and Joshua Hagen
THE BRAIN Michael O'Shea
BRANDING Robert Jones
THE BRICS Andrew F. Cooper
BRITISH CINEMA Charles Barr
THE BRITISH CONSTITUTION Martin Loughlin
THE BRITISH EMPIRE Ashley Jackson
BRITISH POLITICS Tony Wright
BUDDHA Michael Carrithers
BUDDHISM Damien Keown
BUDDHIST ETHICS Damien Keown
BYZANTIUM Peter Sarris

CALVINISM Jon Balserak
ALBERT CAMUS Oliver Gloag
CANADA Donald Wright
CANCER Nicholas James
CAPITALISM James Fulcher
CATHOLICISM Gerald O'Collins
CAUSATION Stephen Mumford and Rani Lill Anjum
THE CELL Terence Allen and Graham Cowling
THE CELTS Barry Cunliffe
CHAOS Leonard Smith
GEOFFREY CHAUCER David Wallace
CHEMISTRY Peter Atkins
CHILD PSYCHOLOGY Usha Goswami
CHILDREN'S LITERATURE Kimberley Reynolds
CHINESE LITERATURE Sabina Knight
CHOICE THEORY Michael Allingham
CHRISTIAN ART Beth Williamson
CHRISTIAN ETHICS D. Stephen Long
CHRISTIANITY Linda Woodhead
CIRCADIAN RHYTHMS Russell Foster and Leon Kreitzman
CITIZENSHIP Richard Bellamy
CITY PLANNING Carl Abbott
CIVIL ENGINEERING David Muir Wood
CLASSICAL LITERATURE William Allan
CLASSICAL MYTHOLOGY Helen Morales
CLASSICS Mary Beard and John Henderson
CLAUSEWITZ Michael Howard
CLIMATE Mark Maslin
CLIMATE CHANGE Mark Maslin
CLINICAL PSYCHOLOGY Susan Llewelyn and Katie Aafjes-van Doorn
COGNITIVE BEHAVIOURAL THERAPY Freda McManus
COGNITIVE NEUROSCIENCE Richard Passingham
THE COLD WAR Robert J. McMahon
COLONIAL AMERICA Alan Taylor
COLONIAL LATIN AMERICAN LITERATURE Rolena Adorno
COMBINATORICS Robin Wilson
COMEDY Matthew Bevis
COMMUNISM Leslie Holmes
COMPARATIVE LITERATURE Ben Hutchinson
COMPETITION AND ANTITRUST LAW Ariel Ezrachi
COMPLEXITY John H. Holland
THE COMPUTER Darrel Ince
COMPUTER SCIENCE Subrata Dasgupta
CONCENTRATION CAMPS Dan Stone
CONDENSED MATTER PHYSICS Ross H. McKenzie
CONFUCIANISM Daniel K. Gardner
THE CONQUISTADORS Matthew Restall and Felipe Fernández-Armesto
CONSCIENCE Paul Strohm
CONSCIOUSNESS Susan Blackmore
CONTEMPORARY ART Julian Stallabrass
CONTEMPORARY FICTION Robert Eaglestone
CONTINENTAL PHILOSOPHY Simon Critchley
COPERNICUS Owen Gingerich
CORAL REEFS Charles Sheppard
CORPORATE SOCIAL RESPONSIBILITY Jeremy Moon
CORRUPTION Leslie Holmes
COSMOLOGY Peter Coles
COUNTRY MUSIC Richard Carlin
CREATIVITY Vlad Glăveanu
CRIME FICTION Richard Bradford
CRIMINAL JUSTICE Julian V. Roberts
CRIMINOLOGY Tim Newburn
CRITICAL THEORY Stephen Eric Bronner
THE CRUSADES Christopher Tyerman
CRYPTOGRAPHY Fred Piper and Sean Murphy
CRYSTALLOGRAPHY A. M. Glazer
THE CULTURAL REVOLUTION Richard Curt Kraus
DADA AND SURREALISM David Hopkins
DANTE Peter Hainsworth and David Robey
DARWIN Jonathan Howard
THE DEAD SEA SCROLLS Timothy H. Lim

DECADENCE David Weir
DECOLONIZATION Dane Kennedy
DEMENTIA Kathleen Taylor
DEMOCRACY Bernard Crick
DEMOGRAPHY Sarah Harper
DEPRESSION Jan Scott and Mary Jane Tacchi
DERRIDA Simon Glendinning
DESCARTES Tom Sorell
DESERTS Nick Middleton
DESIGN John Heskett
DEVELOPMENT Ian Goldin
DEVELOPMENTAL BIOLOGY Lewis Wolpert
THE DEVIL Darren Oldridge
DIASPORA Kevin Kenny
CHARLES DICKENS Jenny Hartley
DICTIONARIES Lynda Mugglestone
DINOSAURS David Norman
DIPLOMATIC HISTORY Joseph M. Siracusa
DOCUMENTARY FILM Patricia Aufderheide
DREAMING J. Allan Hobson
DRUGS Les Iversen
DRUIDS Barry Cunliffe
DYNASTY Jeroen Duindam
DYSLEXIA Margaret J. Snowling
EARLY MUSIC Thomas Forrest Kelly
THE EARTH Martin Redfern
EARTH SYSTEM SCIENCE Tim Lenton
ECOLOGY Jaboury Ghazoul
ECONOMICS Partha Dasgupta
EDUCATION Gary Thomas
EGYPTIAN MYTH Geraldine Pinch
EIGHTEENTH-CENTURY BRITAIN Paul Langford
THE ELEMENTS Philip Ball
EMOTION Dylan Evans
EMPIRE Stephen Howe
EMPLOYMENT LAW David Cabrelli
ENERGY SYSTEMS Nick Jenkins
ENGELS Terrell Carver
ENGINEERING David Blockley
THE ENGLISH LANGUAGE Simon Horobin
ENGLISH LITERATURE Jonathan Bate
THE ENLIGHTENMENT John Robertson
ENTREPRENEURSHIP Paul Westhead and Mike Wright
ENVIRONMENTAL ECONOMICS Stephen Smith
ENVIRONMENTAL ETHICS Robin Attfield
ENVIRONMENTAL LAW Elizabeth Fisher
ENVIRONMENTAL POLITICS Andrew Dobson
ENZYMES Paul Engel
EPICUREANISM Catherine Wilson
EPIDEMIOLOGY Rodolfo Saracci
ETHICS Simon Blackburn
ETHNOMUSICOLOGY Timothy Rice
THE ETRUSCANS Christopher Smith
EUGENICS Philippa Levine
THE EUROPEAN UNION Simon Usherwood and John Pinder
EUROPEAN UNION LAW Anthony Arnull
EVANGELICALISM John G. Stackhouse Jr.
EVIL Luke Russell
EVOLUTION Brian and Deborah Charlesworth
EXISTENTIALISM Thomas Flynn
EXPLORATION Stewart A. Weaver
EXTINCTION Paul B. Wignall
THE EYE Michael Land
FAIRY TALE Marina Warner
FAMILY LAW Jonathan Herring
MICHAEL FARADAY Frank A. J. L. James
FASCISM Kevin Passmore
FASHION Rebecca Arnold
FEDERALISM Mark J. Rozell and Clyde Wilcox
FEMINISM Margaret Walters
FILM Michael Wood
FILM MUSIC Kathryn Kalinak
FILM NOIR James Naremore
FIRE Andrew C. Scott
THE FIRST WORLD WAR Michael Howard
FLUID MECHANICS Eric Lauga
FOLK MUSIC Mark Slobin
FOOD John Krebs
FORENSIC PSYCHOLOGY David Canter

FORENSIC SCIENCE Jim Fraser
FORESTS Jaboury Ghazoul
FOSSILS Keith Thomson
FOUCAULT Gary Gutting
THE FOUNDING FATHERS
R. B. Bernstein
FRACTALS Kenneth Falconer
FREE SPEECH Nigel Warburton
FREE WILL Thomas Pink
FREEMASONRY Andreas Önnerfors
FRENCH LITERATURE John D. Lyons
FRENCH PHILOSOPHY
Stephen Gaukroger and Knox Peden
THE FRENCH REVOLUTION
William Doyle
FREUD Anthony Storr
FUNDAMENTALISM Malise Ruthven
FUNGI Nicholas P. Money
THE FUTURE Jennifer M. Gidley
GALAXIES John Gribbin
GALILEO Stillman Drake
GAME THEORY Ken Binmore
GANDHI Bhikhu Parekh
GARDEN HISTORY Gordon Campbell
GENES Jonathan Slack
GENIUS Andrew Robinson
GENOMICS John Archibald
GEOGRAPHY John Matthews and
David Herbert
GEOLOGY Jan Zalasiewicz
GEOMETRY Maciej Dunajski
GEOPHYSICS William Lowrie
GEOPOLITICS Klaus Dodds
GERMAN LITERATURE Nicholas Boyle
GERMAN PHILOSOPHY
Andrew Bowie
THE GHETTO Bryan Cheyette
GLACIATION David J. A. Evans
GLOBAL CATASTROPHES Bill McGuire
GLOBAL ECONOMIC HISTORY
Robert C. Allen
GLOBAL ISLAM Nile Green
GLOBALIZATION Manfred B. Steger
GOD John Bowker
GÖDEL'S THEOREM A. W. Moore
GOETHE Ritchie Robertson
THE GOTHIC Nick Groom
GOVERNANCE Mark Bevir
GRAVITY Timothy Clifton
THE GREAT DEPRESSION AND
THE NEW DEAL Eric Rauchway
HABEAS CORPUS Amanda L. Tyler
HABERMAS James Gordon Finlayson
THE HABSBURG EMPIRE
Martyn Rady
HAPPINESS Daniel M. Haybron
THE HARLEM RENAISSANCE
Cheryl A. Wall
THE HEBREW BIBLE AS
LITERATURE Tod Linafelt
HEGEL Peter Singer
HEIDEGGER Michael Inwood
THE HELLENISTIC AGE
Peter Thonemann
HEREDITY John Waller
HERMENEUTICS
Jens Zimmermann
HERODOTUS Jennifer T. Roberts
HIEROGLYPHS Penelope Wilson
HINDUISM Kim Knott
HISTORY John H. Arnold
THE HISTORY OF ASTRONOMY
Michael Hoskin
THE HISTORY OF CHEMISTRY
William H. Brock
THE HISTORY OF CHILDHOOD
James Marten
THE HISTORY OF CINEMA
Geoffrey Nowell-Smith
THE HISTORY OF COMPUTING
Doron Swade
THE HISTORY OF LIFE
Michael Benton
THE HISTORY OF MATHEMATICS
Jacqueline Stedall
THE HISTORY OF MEDICINE
William Bynum
THE HISTORY OF PHYSICS
J. L. Heilbron
THE HISTORY OF POLITICAL
THOUGHT Richard Whatmore
THE HISTORY OF TIME
Leofranc Holford-Strevens
HIV AND AIDS Alan Whiteside
HOBBES Richard Tuck
HOLLYWOOD Peter Decherney
THE HOLY ROMAN EMPIRE
Joachim Whaley
HOME Michael Allen Fox
HOMER Barbara Graziosi
HORMONES Martin Luck
HORROR Darryl Jones

HUMAN ANATOMY Leslie Klenerman
HUMAN EVOLUTION Bernard Wood
HUMAN PHYSIOLOGY Jamie A. Davies
HUMAN RESOURCE MANAGEMENT Adrian Wilkinson
HUMAN RIGHTS Andrew Clapham
HUMANISM Stephen Law
HUME James A. Harris
HUMOUR Noël Carroll
THE ICE AGE Jamie Woodward
IDENTITY Florian Coulmas
IDEOLOGY Michael Freeden
THE IMMUNE SYSTEM Paul Klenerman
INDIAN CINEMA Ashish Rajadhyaksha
INDIAN PHILOSOPHY Sue Hamilton
THE INDUSTRIAL REVOLUTION Robert C. Allen
INFECTIOUS DISEASE Marta L. Wayne and Benjamin M. Bolker
INFINITY Ian Stewart
INFORMATION Luciano Floridi
INNOVATION Mark Dodgson and David Gann
INTELLECTUAL PROPERTY Siva Vaidhyanathan
INTELLIGENCE Ian J. Deary
INTERNATIONAL LAW Vaughan Lowe
INTERNATIONAL MIGRATION Khalid Koser
INTERNATIONAL RELATIONS Christian Reus-Smit
INTERNATIONAL SECURITY Christopher S. Browning
INSECTS Simon Leather
IRAN Ali M. Ansari
ISLAM Malise Ruthven
ISLAMIC HISTORY Adam Silverstein
ISLAMIC LAW Mashood A. Baderin
ISOTOPES Rob Ellam
ITALIAN LITERATURE Peter Hainsworth and David Robey
HENRY JAMES Susan L. Mizruchi
JESUS Richard Bauckham
JEWISH HISTORY David N. Myers
JEWISH LITERATURE Ilan Stavans
JOURNALISM Ian Hargreaves
JAMES JOYCE Colin MacCabe
JUDAISM Norman Solomon
JUNG Anthony Stevens
KABBALAH Joseph Dan
KAFKA Ritchie Robertson
KANT Roger Scruton
KEYNES Robert Skidelsky
KIERKEGAARD Patrick Gardiner
KNOWLEDGE Jennifer Nagel
THE KORAN Michael Cook
KOREA Michael J. Seth
LAKES Warwick F. Vincent
LANDSCAPE ARCHITECTURE Ian H. Thompson
LANDSCAPES AND GEOMORPHOLOGY Andrew Goudie and Heather Viles
LANGUAGES Stephen R. Anderson
LATE ANTIQUITY Gillian Clark
LAW Raymond Wacks
THE LAWS OF THERMODYNAMICS Peter Atkins
LEADERSHIP Keith Grint
LEARNING Mark Haselgrove
LEIBNIZ Maria Rosa Antognazza
C. S. LEWIS James Como
LIBERALISM Michael Freeden
LIGHT Ian Walmsley
LINCOLN Allen C. Guelzo
LINGUISTICS Peter Matthews
LITERARY THEORY Jonathan Culler
LOCKE John Dunn
LOGIC Graham Priest
LOVE Ronald de Sousa
MARTIN LUTHER Scott H. Hendrix
MACHIAVELLI Quentin Skinner
MADNESS Andrew Scull
MAGIC Owen Davies
MAGNA CARTA Nicholas Vincent
MAGNETISM Stephen Blundell
MALTHUS Donald Winch
MAMMALS T. S. Kemp
MANAGEMENT John Hendry
NELSON MANDELA Elleke Boehmer
MAO Delia Davin
MARINE BIOLOGY Philip V. Mladenov
MARKETING Kenneth Le Meunier-FitzHugh
THE MARQUIS DE SADE John Phillips
MARTYRDOM Jolyon Mitchell

MARX Peter Singer
MATERIALS Christopher Hall
MATHEMATICAL FINANCE Mark H. A. Davis
MATHEMATICS Timothy Gowers
MATTER Geoff Cottrell
THE MAYA Matthew Restall and Amara Solari
THE MEANING OF LIFE Terry Eagleton
MEASUREMENT David Hand
MEDICAL ETHICS Michael Dunn and Tony Hope
MEDICAL LAW Charles Foster
MEDIEVAL BRITAIN John Gillingham and Ralph A. Griffiths
MEDIEVAL LITERATURE Elaine Treharne
MEDIEVAL PHILOSOPHY John Marenbon
MEMORY Jonathan K. Foster
METAPHYSICS Stephen Mumford
METHODISM William J. Abraham
THE MEXICAN REVOLUTION Alan Knight
MICROBIOLOGY Nicholas P. Money
MICROBIOMES Angela E. Douglas
MICROECONOMICS Avinash Dixit
MICROSCOPY Terence Allen
THE MIDDLE AGES Miri Rubin
MILITARY JUSTICE Eugene R. Fidell
MILITARY STRATEGY Antulio J. Echevarria II
JOHN STUART MILL Gregory Claeys
MINERALS David Vaughan
MIRACLES Yujin Nagasawa
MODERN ARCHITECTURE Adam Sharr
MODERN ART David Cottington
MODERN BRAZIL Anthony W. Pereira
MODERN CHINA Rana Mitter
MODERN DRAMA Kirsten E. Shepherd-Barr
MODERN FRANCE Vanessa R. Schwartz
MODERN INDIA Craig Jeffrey
MODERN IRELAND Senia Pašeta
MODERN ITALY Anna Cento Bull
MODERN JAPAN Christopher Goto-Jones
MODERN LATIN AMERICAN LITERATURE Roberto González Echevarría
MODERN WAR Richard English
MODERNISM Christopher Butler
MOLECULAR BIOLOGY Aysha Divan and Janice A. Royds
MOLECULES Philip Ball
MONASTICISM Stephen J. Davis
THE MONGOLS Morris Rossabi
MONTAIGNE William M. Hamlin
MOONS David A. Rothery
MORMONISM Richard Lyman Bushman
MOUNTAINS Martin F. Price
MUHAMMAD Jonathan A. C. Brown
MULTICULTURALISM Ali Rattansi
MULTILINGUALISM John C. Maher
MUSIC Nicholas Cook
MUSIC AND TECHNOLOGY Mark Katz
MYTH Robert A. Segal
NANOTECHNOLOGY Philip Moriarty
NAPOLEON David A. Bell
THE NAPOLEONIC WARS Mike Rapport
NATIONALISM Steven Grosby
NATIVE AMERICAN LITERATURE Sean Teuton
NAVIGATION Jim Bennett
NAZI GERMANY Jane Caplan
NEGOTIATION Carrie Menkel-Meadow
NEOLIBERALISM Manfred B. Steger and Ravi K. Roy
NETWORKS Guido Caldarelli and Michele Catanzaro
THE NEW TESTAMENT Luke Timothy Johnson
THE NEW TESTAMENT AS LITERATURE Kyle Keefer
NEWTON Robert Iliffe
NIETZSCHE Michael Tanner
NINETEENTH-CENTURY BRITAIN Christopher Harvie and H. C. G. Matthew
THE NORMAN CONQUEST George Garnett
NORTH AMERICAN INDIANS Theda Perdue and Michael D. Green
NORTHERN IRELAND Marc Mulholland
NOTHING Frank Close

NUCLEAR PHYSICS Frank Close
NUCLEAR POWER Maxwell Irvine
NUCLEAR WEAPONS
Joseph M. Siracusa
NUMBER THEORY Robin Wilson
NUMBERS Peter M. Higgins
NUTRITION David A. Bender
OBJECTIVITY Stephen Gaukroger
OCEANS Dorrik Stow
THE OLD TESTAMENT
Michael D. Coogan
THE ORCHESTRA D. Kern Holoman
ORGANIC CHEMISTRY Graham Patrick
ORGANIZATIONS Mary Jo Hatch
ORGANIZED CRIME
Georgios A. Antonopoulos and
Georgios Papanicolaou
ORTHODOX CHRISTIANITY
A. Edward Siecienski
OVID Llewelyn Morgan
PAGANISM Owen Davies
PAKISTAN Pippa Virdee
THE PALESTINIAN-ISRAELI
CONFLICT Martin Bunton
PANDEMICS Christian W. McMillen
PARTICLE PHYSICS Frank Close
PAUL E. P. Sanders
IVAN PAVLOV Daniel P. Todes
PEACE Oliver P. Richmond
PENTECOSTALISM William K. Kay
PERCEPTION Brian Rogers
THE PERIODIC TABLE Eric R. Scerri
PHILOSOPHICAL METHOD
Timothy Williamson
PHILOSOPHY Edward Craig
PHILOSOPHY IN THE ISLAMIC
WORLD Peter Adamson
PHILOSOPHY OF BIOLOGY
Samir Okasha
PHILOSOPHY OF LAW
Raymond Wacks
PHILOSOPHY OF MIND
Barbara Gail Montero
PHILOSOPHY OF PHYSICS
David Wallace
PHILOSOPHY OF SCIENCE
Samir Okasha
PHILOSOPHY OF RELIGION
Tim Bayne
PHOTOGRAPHY Steve Edwards
PHYSICAL CHEMISTRY Peter Atkins
PHYSICS Sidney Perkowitz
PILGRIMAGE Ian Reader
PLAGUE Paul Slack
PLANETARY SYSTEMS
Raymond T. Pierrehumbert
PLANETS David A. Rothery
PLANTS Timothy Walker
PLATE TECTONICS Peter Molnar
PLATO Julia Annas
POETRY Bernard O'Donoghue
POLITICAL PHILOSOPHY David Miller
POLITICS Kenneth Minogue
POLYGAMY Sarah M. S. Pearsall
POPULISM Cas Mudde and
Cristóbal Rovira Kaltwasser
POSTCOLONIALISM Robert J. C. Young
POSTMODERNISM Christopher Butler
POSTSTRUCTURALISM
Catherine Belsey
POVERTY Philip N. Jefferson
PREHISTORY Chris Gosden
PRESOCRATIC PHILOSOPHY
Catherine Osborne
PRIVACY Raymond Wacks
PROBABILITY John Haigh
PROGRESSIVISM Walter Nugent
PROHIBITION W. J. Rorabaugh
PROJECTS Andrew Davies
PROTESTANTISM Mark A. Noll
PSYCHIATRY Tom Burns
PSYCHOANALYSIS Daniel Pick
PSYCHOLOGY Gillian Butler and
Freda McManus
PSYCHOLOGY OF MUSIC
Elizabeth Hellmuth Margulis
PSYCHOPATHY Essi Viding
PSYCHOTHERAPY Tom Burns and
Eva Burns-Lundgren
PUBLIC ADMINISTRATION
Stella Z. Theodoulou and Ravi K. Roy
PUBLIC HEALTH Virginia Berridge
PURITANISM Francis J. Bremer
THE QUAKERS Pink Dandelion
QUANTUM THEORY
John Polkinghorne
RACISM Ali Rattansi
RADIOACTIVITY Claudio Tuniz
RASTAFARI Ennis B. Edmonds
READING Belinda Jack
THE REAGAN REVOLUTION Gil Troy
REALITY Jan Westerhoff

RECONSTRUCTION Allen C. Guelzo
THE REFORMATION Peter Marshall
REFUGEES Gil Loescher
RELATIVITY Russell Stannard
RELIGION Thomas A. Tweed
RELIGION IN AMERICA Timothy Beal
THE RENAISSANCE Jerry Brotton
RENAISSANCE ART Geraldine A. Johnson
RENEWABLE ENERGY Nick Jelley
REPTILES T. S. Kemp
REVOLUTIONS Jack A. Goldstone
RHETORIC Richard Toye
RISK Baruch Fischhoff and John Kadvany
RITUAL Barry Stephenson
RIVERS Nick Middleton
ROBOTICS Alan Winfield
ROCKS Jan Zalasiewicz
ROMAN BRITAIN Peter Salway
THE ROMAN EMPIRE Christopher Kelly
THE ROMAN REPUBLIC David M. Gwynn
ROMANTICISM Michael Ferber
ROUSSEAU Robert Wokler
RUSSELL A. C. Grayling
THE RUSSIAN ECONOMY Richard Connolly
RUSSIAN HISTORY Geoffrey Hosking
RUSSIAN LITERATURE Catriona Kelly
THE RUSSIAN REVOLUTION S. A. Smith
SAINTS Simon Yarrow
SAMURAI Michael Wert
SAVANNAS Peter A. Furley
SCEPTICISM Duncan Pritchard
SCHIZOPHRENIA Chris Frith and Eve Johnstone
SCHOPENHAUER Christopher Janaway
SCIENCE AND RELIGION Thomas Dixon and Adam R. Shapiro
SCIENCE FICTION David Seed
THE SCIENTIFIC REVOLUTION Lawrence M. Principe
SCOTLAND Rab Houston
SECULARISM Andrew Copson
SEXUAL SELECTION Marlene Zuk and Leigh W. Simmons
SEXUALITY Véronique Mottier
WILLIAM SHAKESPEARE Stanley Wells
SHAKESPEARE'S COMEDIES Bart van Es
SHAKESPEARE'S SONNETS AND POEMS Jonathan F. S. Post
SHAKESPEARE'S TRAGEDIES Stanley Wells
GEORGE BERNARD SHAW Christopher Wixson
MARY SHELLEY Charlotte Gordon
THE SHORT STORY Andrew Kahn
SIKHISM Eleanor Nesbitt
SILENT FILM Donna Kornhaber
THE SILK ROAD James A. Millward
SLANG Jonathon Green
SLEEP Steven W. Lockley and Russell G. Foster
SMELL Matthew Cobb
ADAM SMITH Christopher J. Berry
SOCIAL AND CULTURAL ANTHROPOLOGY John Monaghan and Peter Just
SOCIAL PSYCHOLOGY Richard J. Crisp
SOCIAL WORK Sally Holland and Jonathan Scourfield
SOCIALISM Michael Newman
SOCIOLINGUISTICS John Edwards
SOCIOLOGY Steve Bruce
SOCRATES C. C. W. Taylor
SOFT MATTER Tom McLeish
SOUND Mike Goldsmith
SOUTHEAST ASIA James R. Rush
THE SOVIET UNION Stephen Lovell
THE SPANISH CIVIL WAR Helen Graham
SPANISH LITERATURE Jo Labanyi
THE SPARTANS Andrew J. Bayliss
SPINOZA Roger Scruton
SPIRITUALITY Philip Sheldrake
SPORT Mike Cronin
STARS Andrew King
STATISTICS David J. Hand
STEM CELLS Jonathan Slack
STOICISM Brad Inwood
STRUCTURAL ENGINEERING David Blockley
STUART BRITAIN John Morrill
SUBURBS Carl Abbott
THE SUN Philip Judge
SUPERCONDUCTIVITY Stephen Blundell
SUPERSTITION Stuart Vyse

SYMMETRY Ian Stewart
SYNAESTHESIA Julia Simner
SYNTHETIC BIOLOGY Jamie A. Davies
SYSTEMS BIOLOGY Eberhard O. Voit
TAXATION Stephen Smith
TEETH Peter S. Ungar
TELESCOPES Geoff Cottrell
TERRORISM Charles Townshend
THEATRE Marvin Carlson
THEOLOGY David F. Ford
THINKING AND REASONING Jonathan St B. T. Evans
THOUGHT Tim Bayne
TIBETAN BUDDHISM Matthew T. Kapstein
TIDES David George Bowers and Emyr Martyn Roberts
TIME Jenann Ismael
TOCQUEVILLE Harvey C. Mansfield
LEO TOLSTOY Liza Knapp
TOPOLOGY Richard Earl
TRAGEDY Adrian Poole
TRANSLATION Matthew Reynolds
THE TREATY OF VERSAILLES Michael S. Neiberg
TRIGONOMETRY Glen Van Brummelen
THE TROJAN WAR Eric H. Cline
TRUST Katherine Hawley
THE TUDORS John Guy
TWENTIETH-CENTURY BRITAIN Kenneth O. Morgan
TYPOGRAPHY Paul Luna
THE UNITED NATIONS Jussi M. Hanhimäki
UNIVERSITIES AND COLLEGES David Palfreyman and Paul Temple
THE U.S. CIVIL WAR Louis P. Masur
THE U.S. CONGRESS Donald A. Ritchie
THE U.S. CONSTITUTION David J. Bodenhamer
THE U.S. SUPREME COURT Linda Greenhouse
UTILITARIANISM Katarzyna de Lazari-Radek and Peter Singer
UTOPIANISM Lyman Tower Sargent
VATICAN II Shaun Blanchard and Stephen Bullivant
VETERINARY SCIENCE James Yeates
THE VIKINGS Julian D. Richards
VIOLENCE Philip Dwyer
THE VIRGIN MARY Mary Joan Winn Leith
THE VIRTUES Craig A. Boyd and Kevin Timpe
VIRUSES Dorothy H. Crawford
VOLCANOES Michael J. Branney and Jan Zalasiewicz
VOLTAIRE Nicholas Cronk
WAR AND RELIGION Jolyon Mitchell and Joshua Rey
WAR AND TECHNOLOGY Alex Roland
WATER John Finney
WAVES Mike Goldsmith
WEATHER Storm Dunlop
THE WELFARE STATE David Garland
WITCHCRAFT Malcolm Gaskill
WITTGENSTEIN A. C. Grayling
WORK Stephen Fineman
WORLD MUSIC Philip Bohlman
WORLD MYTHOLOGY David Leeming
THE WORLD TRADE ORGANIZATION Amrita Narlikar
WORLD WAR II Gerhard L. Weinberg
WRITING AND SCRIPT Andrew Robinson
ZIONISM Michael Stanislawski
ÉMILE ZOLA Brian Nelson

Available soon:

OBSERVATIONAL ASTRONOMY Geoff Cottrell
MATHEMATICAL ANALYSIS Richard Earl
HISTORY OF EMOTIONS Thomas Dixon

For more information visit our website

www.oup.com/vsi/

Ross H. McKenzie

CONDENSED MATTER PHYSICS

A Very Short Introduction

OXFORD
UNIVERSITY PRESS

Great Clarendon Street, Oxford, OX2 6DP,
United Kingdom

Oxford University Press is a department of the University of Oxford.
It furthers the University's objective of excellence in research, scholarship,
and education by publishing worldwide. Oxford is a registered trade mark of
Oxford University Press in the UK and in certain other countries

First edition published in 2023

Published in the United States of America by Oxford University Press
198 Madison Avenue, New York, NY 10016, United States of America

British Library Cataloguing in Publication Data
Data available

Library of Congress Control Number: 2022943040

ISBN 978–0–19–884542–3

Printed and bound by
CPI Group (UK) Ltd, Croydon, CR0 4YY

Contents

Preface

Condensed matter physics is one of the largest and most vibrant sub-fields of physics. It is a rich source of ideas, concepts, and techniques that have cross-fertilized with not just other sub-fields of physics, such as elementary particles, but also other scientific disciplines (chemistry, biology, computer science, sociology, economics) and engineering (electrical, materials, chemical). In the past 30 years, the Nobel Prize in Physics has been awarded 13 times for work on condensed matter. In the past 20 years, eight condensed matter physicists have received the Nobel Prize in Chemistry.

Condensed matter physics is all about emergence; the whole is greater than the sum of the parts. A system composed of many interacting parts (atoms) has properties (a state of matter) that are qualitatively different from the properties of the individual parts. This commonality is why condensed matter physics has had such a fruitful cross-fertilization with other sciences. It is arguably the field with the greatest success at understanding emergent phenomena in complex systems, particularly at the quantitative level. This is not because condensed matter physicists are smarter than sociologists, economists, or neuroscientists. It is because the materials we study are much 'simpler' than societies, economies, and brains.

As a research field, condensed matter physics is not just defined by the objects that it studies (states of matter in materials), but rather by a particular *approach* to the study of these objects. The aim is to address fundamental questions and to find unifying concepts that can be used to describe and understand a wide range of phenomena in materials that are chemically and structurally diverse. My aim has not been to write a popular summary of the main topics covered in a textbook. Rather this is an essay that I hope brings joy and appreciation and a desire to learn more. I have chosen to emphasize concepts rather than trying to explain technical details.

This VSI is shaped by an important intellectual legacy that should be acknowledged and only became apparent to me when most of the book was written. In his recent biography, *Mind over Matter: Philip Anderson and the Physics of the Very Many*, Andrew Zangwill states, 'more than any other twentieth-century physicist, he [Anderson] transformed the patchwork of ideas and techniques formerly called solid-state physics into the deep, subtle, and intellectually coherent discipline known today as *condensed matter physics*'. I hope that what I have written does honour this legacy of Phil Anderson (1923–2020). I also hope that I can convince you that condensed matter physics is significant, exciting, beautiful, and profound.

Acknowledgements

I thank my editor at OUP, Latha Menon, for her support, constructive feedback, patience, and diligent editing. I thank Stephen Blundell for introducing me to Latha and supporting this project.

Several institutions provided opportunities and hospitality, including University of Queensland, Trinity College Queensland, and Wolfson College (Oxford).

I am grateful for generous and substantial financial support over many years for other projects that produced relevant ideas and experiences that helped shape the book. I acknowledge the University of Queensland, Australian Research Council, and John Templeton Foundation.

I have worked with many students, postdocs, and collaborators who have taught me much, some of which has contributed to this book.

The book was only possible because of the education and employment that I received over four decades. The following people believed in me and supported me at critical junctions: Hans Buchdahl, Jim Sauls, John Wilkins, Jaan Oitmaa, Gerard

Milburn, and Andrew Taylor. I am very grateful for the opportunities that they provided access to.

I have received encouragement, substantial comments, and feedback on drafts of all parts of the text. This has led to many rewrites that have improved the clarity, accuracy, and interest of the manuscript. For this I thank the members of the 'Holy' Scribblers writers collective, Janay Garrick from Understory Creatives, Luke McKenzie, Stephen Ney, Andrew Zangwill, and many commenters on my blog.

Finally, I am very grateful to my wife, Robin, and children, Luke and Michelle, for their interest and support.

List of illustrations

Chapter 1
What is condensed matter physics?

Every day we encounter a diversity of materials: liquids, glass, ceramics, metals, crystals, magnets, plastics, semiconductors, foams…These materials look and feel different from one another. Their physical properties vary significantly: are they soft and squishy or hard and rigid? Shiny, black, or colourful? Do they absorb heat easily? Do they conduct electricity? The distinct physical properties of different materials are central to their use in technologies around us: smartphones, alloys, semiconductor chips, computer memories, cooking pots, magnets in MRI machines, LEDs in solid-state lighting, and fibre optic cables. Consequently, the science of materials attracts researchers in a wide range of disciplines: physics, chemistry, biology, mathematics, and the varieties of engineering (electrical, chemical, mechanical, material…). But why do different materials have different physical properties?

There are more than 100 different types of atoms, or chemical elements, in the universe. Any material is composed of a specific collection of different atoms, and they are arranged in a particular spatial pattern within the material. A central question is:

> How are the physical properties of a material related to the properties of the atoms from which the material is made?

Diamond versus graphite

To better understand how this question is addressed by condensed matter physicists let's consider a concrete example, that of graphite and diamond. While you will find very cheap graphite in lead pencils, you will find diamonds with a hefty price tag upwards of thousands of dollars depending on the 4 Cs (carat, colour, clarity, and cut). But both graphite and diamond are composed solely of carbon atoms. So why do they look and feel so different?

Let's make some observations about the physical *properties* of graphite versus the properties of diamond. They are both solid. Graphite is common, black, soft, and conducts electricity and heat moderately well. In contrast, diamond is rare, transparent, hard, and conducts electricity and heat very poorly.

What is the origin of these distinctly different physical properties of graphite and diamond? How can they be so different if they are both made of the same atoms? This is where we need a special 'microscope' to zoom in and look at the diamond and graphite at very high magnifications, that is, at the level of seeing the individual carbon atoms of which the materials are made. This can be done by shining a beam of X-rays on each of the materials. The X-ray 'microscope' reveals that the carbon atoms have a distinctly different geometric arrangement in diamond and in graphite. But first, let's look at the atomic scale structure of graphene (Figure 1), a single sheet of carbon atoms.

Now imagine taking a bunch of graphene sheets and stacking them one on top of each other. This is the structure of graphite (Figure 2). At the atomic scale graphite is composed of layers of graphene. In each layer the atoms are arranged in a honeycomb-like sheet that consists of an infinite array of hexagons and each carbon atom is next to three neighbouring carbon atoms. In diamond, every carbon atom is next to four other carbon atoms

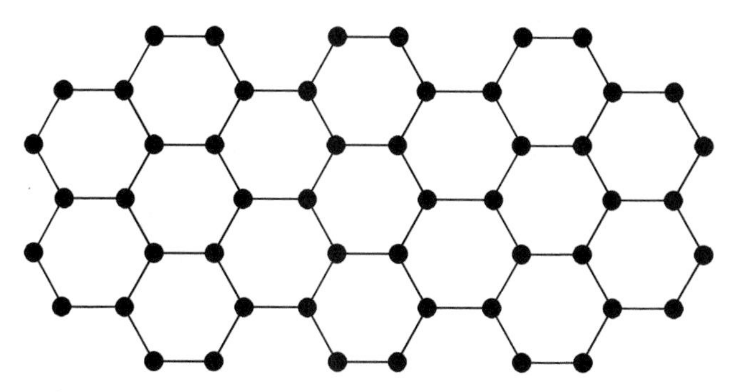

1. The microscopic structure of graphene. Each solid circle represents a single carbon atom and each solid line a chemical bond between two neighbouring atoms.

and there are no sheets of atoms. The two atomic arrangements are *qualitatively different*. In diamond the network of chemical bonds extends in three directions, whereas in graphite it only extends in two directions, that is, parallel to the sheets.

The two different structures at the atomic (microscopic) scale led to the difference in physical properties of graphite and diamond at the macroscopic scale. Graphite is soft because it is easy to make the different sheets of atoms slide relative to one another. When you write with a pencil this is how the 'lead' in your pencil easily slides off onto a piece of paper. Diamond is hard because there are no such layers; the carbon atoms form a strong interpenetrating network and so it is difficult to move them relative to one another. Explaining the differences in colour and conductivity between diamond and graphite is more difficult. It requires the use of quantum theory to explain how electrons move through the infinite network of atoms within each of the two solid states.

States of matter

Diamond and graphite are distinct solid states of carbon. They have *qualitatively different* physical properties, at both the

Graphite vs Diamond

Graphite *Diamond*

Dull, opaque, soft, common *Brilliant, transparent, hand, rare*

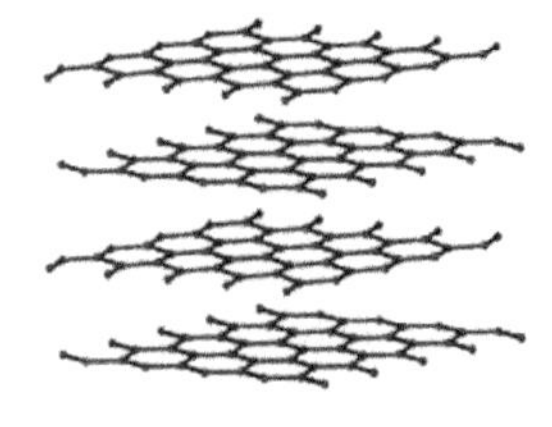

2. A side-by-side comparison of the different properties of graphite and diamond. The upper figures show the macroscopic physical properties. The lower figures show the microscopic properties: the geometric arrangement of the carbon atoms within each of the materials.

microscopic and the macroscopic scale. Condensed matter physics is all about states of matter. In science classes at school, you were probably taught that there are only three states of matter: solid, liquid, and gas. Like other things you were told in school, this is incorrect. There are endless, unlimited, distinct states of matter.

Consider the 'liquid crystals' that are the basis of LCDs (Liquid Crystal Displays) in the screens of televisions, computers, and smartphones. How can something be both a liquid *and* a crystal? A liquid crystal is a distinct state of matter. Solids can be found in many different states. We have already seen that there are two different solid states of carbon: graphite and diamond.

In everyday life ice means simply solid water. But there are in fact *18* different solid states of water, depending on the temperature of the water and the pressure that is applied to the ice. In each of these 18 states there is a unique spatial arrangement of the water molecules and there are qualitative differences in the physical properties of the different solid states. Welcome to the world of condensed matter.

Condensed matter physics is concerned with characterizing and understanding all the different states of matter that can exist. Each state of matter has unique physical properties. Qualitative differences distinguish the different states of matter. These different states are called condensed states (or phases) of matter. The word 'condensed' is used here in the same sense as when we say that steam condenses into liquid water. Generally, as the temperature is lowered, a material can condense into a new state of matter. At high temperatures carbon is a gas and when the temperature is lowered it condenses into a liquid. Further cooling causes liquid carbon to condense into graphite. At high pressures when graphite is cooled it can condense into diamond.

Materials are studied by physicists, chemists, and engineers, and the questions, focus, goals, and techniques of researchers from these different disciplines can be quite different. The focus of condensed matter physics is on states of matter. Condensed matter physics as a research field is not just defined by the objects that it studies (states of matter in materials), but rather by a particular *approach* to the study of these objects. The aim is to address fundamental questions and to find unifying concepts and organizing principles to understand a wide range of phenomena in materials that are chemically and structurally diverse.

The central question of condensed matter physics

Generally, condensed matter physicists grapple with one question. Because it is so important, I state the question in three different ways. How do *macroscopic* properties *emerge* from *microscopic* properties? How do the properties of a state of matter *emerge* from the properties of the atoms in the material and their interactions? How do the many atoms in a material interact with one another to *collectively* produce a particular property of the material?

Answering this question requires determining three things about a material: the *macroscopic* properties; the *microscopic* properties; and how they are related. Determining any of these is a scientific challenge. Nevertheless, condensed matter physicists, often in collaboration with scientists from other fields, have creatively and painstakingly developed a wide range of methods to determine them.

Superconductivity

To me, and many other condensed matter physicists, the most interesting state of matter is superconductivity. In this state, a metal can conduct electricity perfectly; without generating any heat. The discovery of superconductivity in 1911 presented a considerable intellectual challenge: what is the origin of this new state of matter? How do the electrons in the metal interact with one another to produce superconductivity? Many of the greatest theoretical physicists of the 20th century took up this challenge but failed: Albert Einstein, Niels Bohr, Wolfgang Pauli, Richard Feynman, and Lev Landau to name a few. The puzzle was only solved 46 years after the discovery by John Bardeen, Leon Cooper, and Robert Schrieffer, who presented what we know today as the BCS theory. Their theory has subsequently been used to understand not just superconductivity, but also liquid helium, nuclear physics, neutron stars in astrophysics, and the Higgs boson. The study of superconductivity is an ongoing story, with many new classes of materials being discovered that show the behaviour. However, it normally occurs at extremely low temperatures, close to absolute zero (–273 °C), which limits its practical applications. A huge discovery, made in 1986, was of a family of ceramic materials that can superconduct at much higher temperatures than others. They just need liquid nitrogen to cool them down. This still gives hope of finding the 'holy grail': a material that can superconduct at room temperature. But more on that later.

I find superconductivity even more interesting when considering quantum effects. By 1930 it was widely accepted that quantum theory, in all its strangeness, describes the atomic world of electrons and protons, but does not show itself in the everyday world of what we can see and touch. You cannot be in two places at the same time. Your cat is either dead or alive. However, condensed matter physicists have shown that the boundary between the atomic and macroscopic worlds is not so clear cut. A piece of superconducting metal can take on weird quantum properties, just like a single atom. But more on that later, too.

Forerunners of condensed matter physics

In terms of institutional structures condensed matter physics became more defined in 1978 when the American Physical Society decided to change the name of its Division of Solid State Physics to the Division of Condensed Matter Physics. Today, it hosts annually the largest physics conference in the world, attracting more than 10,000 attendees. But long before 1978 scientists had been busy studying condensed states of matter. Two scientists, Heike Kamerlingh Onnes and Lev Landau, began laying the foundations of the field.

Heike Kamerlingh Onnes might be the first condensed matter physicist, because he embodied the intellectual orientation and integrated multifaceted approach that is at the heart of condensed matter physics today. Kamerlingh Onnes was driven by a desire to answer fundamental questions such as, can any gas become liquid? Is there a universal relationship between the density, pressure, and temperature of a gas? How are gas–liquid transitions related to interactions between the constituent molecules in a material? At very low temperatures is the electrical conductivity of a pure metal zero, finite, or infinite? To help design experiments and understand their results, Kamerlingh Onnes sought out theoretical advice from his colleague in Leiden, in the Netherlands, Johannes van der Waals.

To work towards answers for such fundamental questions, Kamerlingh Onnes played the long game. He spent years developing and improving experimental methods and techniques whether glass blowing, sample purification, or building vacuum pumps. He realized that this approach required a large team of technicians, each with particular expertise, and that teamwork was important. The motto of Kamerlingh Onnes's laboratory was *Door meten tot weten*, 'Through measurement to knowledge.'

Kamerlingh Onnes is credited with the discovery of superconductivity in 1911. Crucial to this discovery was an earlier landmark achievement. Kamerlingh Onnes's group were the first to cool helium gas to low enough temperatures that it would become liquid. At the time all other known gases had been liquefied, but not helium. It wasn't clear whether the problem was that scientists could not get the gas cold enough or whether helium was always a gas. In 1908, Kamerlingh Onnes's group observed that helium did become liquid at a temperature of −269 °C. This discovery was of both fundamental importance and practical significance. Liquid helium later became extremely useful to cool materials and scientific instruments to very low temperatures. Today, many large universities have storage tanks that supply science and engineering departments with the liquid helium that they need for their research programmes.

Lev Landau was arguably the first condensed matter *theorist*, that is a theoretical physicist who developed mathematical theories to

3. Landau and Kapitsa in 1948.

describe the results of experiments on condensed states of matter. In 1937 Landau published three papers that marked the beginning of theoretical condensed matter physics. In these papers, he provided a concrete methodology to classify different states of matter. Landau lived in the Soviet Union and his 1937 papers were almost the last he wrote. In 1938 he was arrested and imprisoned for comparing Stalinism to Nazism. Pyotr Kapitsa, the director of the research institute where Landau worked, wrote to Stalin requesting Landau's release, but to no avail. After a year, Kapitsa wrote to Vyacheslav Molotov (of cocktail fame), Stalin's protégé, arguing that Landau was indispensable to help him 'clear up one of the most puzzling areas in modern physics'. This letter did the trick and in the year following his release Landau developed a theory to explain Kapitsa's experimental results on *superfluidity* (another intriguing state of matter) in liquid helium. Both Landau and Kapitsa received Nobel Prizes for their work on superfluid helium (Figure 3).

Emergence

Condensed matter physics can be contrasted with two other 'fundamental' sub-fields of physics: cosmology and elementary particle physics. Cosmology is concerned with the extremely *large*: the scale and history of the whole universe. Elementary particle physics is concerned with the extremely *small*: finding the elementary building blocks of all matter: quarks, electrons, Higgs bosons, and the forces between these fundamental particles. In contrast, condensed matter physics is concerned with the extremely *complex*: how large numbers of atoms or molecules behave when they interact with one another. Tackling this great complexity involves the challenge of connecting microscopic and macroscopic properties, which, in condensed matter physics, is all about *emergence*, one of the most important concepts in science.

Emergence refers to the observation that the whole is greater than the sum of the parts. The whole is *qualitatively different*. A system composed of many interacting parts has properties that often cannot be anticipated from a knowledge of the properties of the individual parts. This concept describes many phenomena at the heart of not only condensed matter physics but also chemistry, biology, psychology, economics, and sociology. For example, a gene is a portion of a very long DNA molecule. The gene has no personality. Yet your appearance and personality emerge from the complex interaction of genes with one another and with proteins in your body. A single neuron does not have thoughts but the trillions of neurons in your brain act together to produce thoughts. A piece of solid gold is shiny. But it makes no sense to say a single gold atom is shiny. Shininess is a property of a collection of gold atoms. Water is wet. But it makes no sense to say that a single water molecule is wet. Wetness is a property of many water molecules that are in the liquid state. Similarly, the macroscopic properties that distinguish graphite and diamond (e.g. soft vs hard, black vs transparent) are emergent. Both diamond and graphite have identical constituents: carbon atoms. Yet they interact in different ways to collectively produce properties that the individual carbon atoms do not have.

Most macroscopic properties are emergent, that is, they are irreducible. Emergent properties are properties of the whole system but not of the individual components of the system. Characteristics of emergent properties include surprising discoveries, the difficulty of making theoretical predictions, the qualitative difference between microscopic and macroscopic properties, and universality. Universality here means that systems with different microscopic details (such as chemical composition) can have the same macroscopic properties.

Another surprise

Everything discussed so far about graphite and diamond has been known and understood since about 1930. But condensed matter physics continues to be full of surprises, and the story of solid carbon continues to unfold. It involves new geometries: tubes, spheres, and sheets of carbon. Looking back at the structure of graphene shown in Figure 1, consider taking a strip of carbon atoms and wrapping up the sheet into a tube. This is known as a carbon nanotube because the diameter of the tube is on the scale of a few nanometres (one billionth of a millimetre). Nanotubes were first made in the laboratory in 1991, and they turned out to be conducting wires with some remarkable properties.

Now consider cutting out a small disk in the graphene sheet and wrapping it up into the shape of a sphere. It turns out that one very stable type of sphere is a collection of 60 carbon atoms with the exact same spatial pattern as a soccer ball. This molecule is known as a fullerene or a buckyball, in honour of the architect

Buckminster Fuller, who popularized structures with some similarities, called geodesic domes. In 1985 a group of chemists were performing experiments to try to understand the formation of molecules of long carbon chains in interstellar space. Unexpectedly, they found that they had made fullerenes in the laboratory. In 1991 condensed matter physicists were excited when they discovered that if a solid chemical compound was made from buckyballs and potassium it was superconducting, and at a temperature much higher than that for simple metals such as tin and lead.

But what about graphene itself, single sheets of carbon? It was not produced in a laboratory until 2004, by a team led by Andre Geim at the University of Manchester in England, and under somewhat unusual circumstances. After a demanding week, Geim's research group liked to try some physics experiments that were fun and slightly crazy on a Friday night. Crazy could mean silly or extremely difficult. This also encouraged creativity in the research group. Most of the experiments they tried did not work. One experiment that did work and became an internet sensation was levitating a frog inside a powerful electromagnet.

One Friday night, Geim wondered whether it was possible to take some graphite and peel off a single layer of carbon atoms. What he proposed doing next is surprisingly simple, particularly in a laboratory equipped with state-of-the-art scientific equipment: attach a piece of sticky tape to graphite, say from a pencil, and then pull it off, hopefully peeling off a single layer of carbon from the pencil in the process. Then take a look at the tape under a high-powered microscope. When the team tried this, to their surprise, they found single layers of carbon: they had made graphene. Sometimes crazy ideas do work out.

Working together with Konstantin Novoselov and other group members, Geim was able to make electrical circuits using the graphene and found that it conducted electricity reasonably well.

The electrons in graphene are responsible for its electrical conductivity. They behave in a manner characteristic of light particles (photons, which have no mass) rather than isolated electrons (which have mass).

The discovery of graphene was so revolutionary that Geim and Novoselov were awarded the Nobel Prize in Physics in 2010. But the story is still not over. In 2016, scientists took two layers of graphene and placed one on top of the other, but at a slightly different orientation, so that they were twisted relative to one another by an angle of just one degree. This arrangement was motivated by theoretical predictions that such a system would conduct electricity in a manner different from graphene. Then, in 2018, these systems were found under certain conditions to be superconducting. Furthermore, in an electrical circuit like in a transistor it was possible to switch the material between three different states of matter: superconductor, metal, and insulator.

The next chapter discusses some of the many known distinct states of matter, and how condensed matter physicists distinguish these states from one another.

Chapter 2
A multitude of states of matter

Classifying objects, people, and societies requires making *qualitative* distinctions. One book is easy to understand, and another is hard. One person is kind, and another is mean. One society is egalitarian, and another is not. Justifying such qualitative distinctions is hard. Not everyone will agree. Are there definitive criteria to justify a particular quality? Some claim they can quantify qualities such as these but that is contentious. In contrast, in condensed matter physics it is possible to give objective criteria that distinguish different states of matter. A state can only exist under specific external conditions, including defined ranges of parameters such as temperature and pressure. This chapter describes the clear signatures of transitions between different states that are observed as these parameters are varied. Some of the many known states of matter will be introduced including superconductors, superfluids, and magnets. On the way we will learn about 'dry ice', how to convert graphite into diamond, and how freeze-dried food is made.

Abrupt changes in properties

If you put some ice cubes in one empty glass and water in another, the ice does not change its shape, whereas water takes the shape of the glass. Solids are rigid and liquids are not. The distinct change from one state to another can be detected by observing an abrupt change or discontinuity in physical properties. For example, ice (solid water) has a different density from liquid water. This is evident because ice floats. The solid state of water has a lower density than the liquid state. To put it another way, water expands when it freezes. That's why water pipes can burst if they freeze in cold weather.

A transition between two distinct states of matter is an example of a *tipping point*: a small change in a system variable can produce large changes in the system. For example, changing the temperature of water from plus 1 °C to minus 1 °C can produce a qualitative change in system properties. The water changes from liquid to solid. Tipping points occur in a wide range of physical, biological, and social systems. Examples include a stock market crash, the outbreak of an epidemic, and the operation of a room thermostat. Tipping points show that *quantitative differences can become qualitative differences*. This concept has a long and controversial heritage. In the 19th century, using the example of transitions between distinct states of matter, the philosopher Georg Friedrich Hegel proposed the first of his three 'laws of dialectics': 'The law of transformation of quantity into quality and *vice versa*.' This 'law' was promoted by Friedrich Engels and was claimed to be part of the scientific basis of Marxist-Leninist ideology, including claims about the qualitative difference between workers and capitalists, the instability of capitalism, and the inevitability of political revolution. Now, back to the less controversial topic of condensed matter.

How does a scientist know if they have discovered a new state of matter? One method to detect the transition between states is to measure how the temperature of the material changes as it is cooled or heated. Put a thermometer in a pot of boiling water on a stove and you will see that the temperature stays fixed at 100 °C, even though heat is being added. The temperature also does not change when heat is added to ice to make it melt. The temperature stays fixed at 0 °C while it melts. Suppose you take a fixed mass of

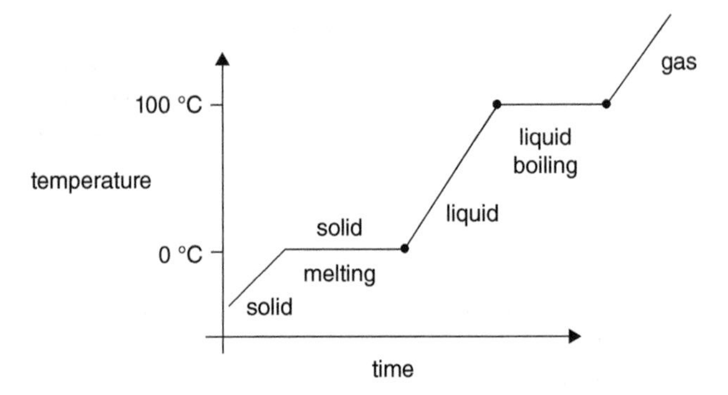

4. Temperature of a fixed mass of water versus time when a constant source of heat is added to the water. The transition between each of the states (solid, liquid, gas) is clearly defined by the flat parts of the graph.

liquid water, place a thermometer in it, and then freeze the water. Then add a constant source of heat. The ice will eventually melt, and the liquid will then heat up and eventually boil. If you measure the temperature of the water as a function of time and graph the results you will get something similar to Figure 4. Heating the block of ice leads to the temperature not changing when the ice is melting and when the liquid water is boiling. The main thing to note is that the flat parts of the curve clearly define the transition from solid to liquid and from liquid to gas. In particular, the state changes are associated with an abrupt change in the slope of the graph. State changes are also called phase transitions. I stress that 'transition' is not used in the sense of a gradual change. Rather, the transition is sudden. State changes are associated with abrupt changes in physical properties.

If this experiment is repeated at a different air pressure, the shape of the curve is the same as in Figure 4, except that the melting and boiling temperatures are different. On the top of Mount Everest, the air pressure is only one-third of that at sea level, and water boils at only 71 °C. So don't expect a piping hot cup of tea to celebrate climbing to the top! Furthermore, if one does a similar experiment for almost any substance that is undergoing a phase transition, one will see qualitatively similar behaviour. There are abrupt changes in the slope of the graph of temperature versus time. This is true not just for solid–liquid transitions but solid–solid transitions, such as between graphite and diamond, and the magnetic and superconducting transitions to be discussed below.

Similarly, if the volume of a fixed mass of water is measured over time when heat is added, there are changes in the slope of the volume versus time curve when the state changes. The solid, liquid, and gas states of a fixed mass of water have different volumes. Again, there are clear measurable signatures of a change in state.

Many new states of matter have been discovered in this manner, by taking a material and cooling it down to lower and lower temperatures and looking for abrupt changes in the physical properties of the material. Liquid crystals, superconductivity, and the superfluid states of liquid helium were all discovered in this manner.

Temperature scales

Temperature is relevant to everyday life: the weather, home heating, cooking food, and storing food. The two most used temperature scales are degrees Celsius (°C) and degrees Fahrenheit (°F). The Celsius scale is convenient because water freezes at 0 °C and boils at 100 °C. The corresponding temperatures in Fahrenheit are 32 °F and 212 °F, respectively. Temperature is central to condensed matter physics. Changing the temperature of a material can induce a transition from one state of matter to another. Measuring how the properties of a state change as the temperature is varied can provide insights into the nature of the state.

A scientifically fundamental temperature scale is the Kelvin scale or *absolute temperature*. This scale emerged from the field of thermodynamics, which is concerned with the relationship between heat, energy, and work. A fundamental law of nature, known as the third law of thermodynamics, is that it is impossible to cool any substance below a temperature of −273.15 °C. At this temperature all the random movement of atoms due to heat ceases. This is known as *absolute zero* and defined as 0 Kelvin. The unit of temperature is a Kelvin, denoted K, and a change of one K corresponds to a change in the Celsius scale of one degree. Water freezes at a temperature of 273.15 K and boils at 373.15 K.

The absolute temperature scale is convenient in condensed matter physics because many interesting states of matter, such as superconductors and superfluids, occur at temperatures less than a few Kelvin. Some even occur at less than one-thousandth of a Kelvin.

Superconductors

Although metals are good conductors of electricity, they are not perfect conductors. Passing an electrical current through a wire generates heat. This heat can lead to electrical fires. Traditional light globes pass current through a very thin wire that generates so much heat that it emits light. This heating is also the basis of old-style electrical fuses, which consist of a very thin wire that melts if there is an electrical short and the current gets too large, breaking the electrical circuit.

If the temperature is lowered, metals become better conductors. But what happens as the temperature of a metal approaches absolute zero? At the beginning of the 20th century there were three different theoretical predictions: the conductivity would become zero, infinite, or be finite. Having liquefied helium, Kamerlingh Onnes was able to address this fundamental question. His group prepared purified samples of gold and mercury and measured how well they conducted electricity, as the temperature was lowered down to one Kelvin. For gold, the electrical resistance steadily decreased and then did not change below about 10 K. But the behaviour of the mercury was completely unexpected. As the temperature decreased, the resistance steadily decreased and then suddenly, at about 4 K, the resistance dramatically dropped to zero. This was first observed by a graduate student, Gilles Holst. The mercury had become a *perfect* electrical conductor. Indeed, a battery was connected to a loop of mercury wire to produce an electrical current, and then the battery was disconnected. The current just kept going and going. In the 1960s this experiment was repeated, and extremely accurate measurements observed no decay of the electrical current over the course of two months. The scientists estimated that the superconducting current would last for at least 100,000 years.

If you have ever had an MRI in a hospital you have benefited from superconductivity. MRI (Magnetic Resonance Imaging) machines use large magnetic fields that are produced by hundreds of kilometres of superconducting wire wound into a coil. An electrical current is put in the wire from an external power supply which is then disconnected. The current and magnetic field can then exist for days without putting any additional electrical energy into the system. In contrast, a regular electromagnet, such as one built with copper wire, requires a continual input of electrical energy. Furthermore, the wires would melt because of the heat generated by the large electrical currents passing through them. In contrast, superconducting magnets do not have this problem because there is no electrical resistance in the wires to generate any heat. The Large Hadron Collider in Europe is the largest and most powerful particle accelerator in the world. It is at the forefront of research on elementary particles, and where the Higgs boson was first observed. A ring of superconducting magnets, arranged in a circle with a diameter of 27 kilometres, direct beams of particles moving at close to the speed of light, before the particles collide with one another. Superconducting magnets are also widely used in chemistry and biochemistry as they are essential to two scientific instruments that can help identify molecules and find their structure: mass spectrometers, which measure the mass of molecules, and NMR (Nuclear Magnetic Resonance) spectrometers.

The perfect conductivity of the superconducting state is not the only property that shows it is a state of matter distinct from the normal conducting state of a regular metal. A second characteristic property is that a magnetic field cannot exist inside a superconductor. This surprising property was discovered in 1933 by Walther Meissner and Robert Ochsenfeld. The fact that magnetic fields cannot penetrate a superconductor means that if a magnet is placed above a superconductor the magnet will be repelled by the superconductor. This makes it possible to levitate large objects. Japanese scientists presented a dramatic demonstration of this effect in the 1990s. A sumo wrestler stood on a large magnetic plate on top of a large superconductor and the wrestler and plate floated. A more practical application is high speed levitating trains.

Superfluids

The specific heat capacity of a substance is a measure of the amount of heat energy that needs to be added to the substance to produce an increase in temperature. Willem Keesom was a former student of Kamerlingh Onnes in Leiden. In 1927, Keesom made a discovery concerning the specific heat capacity of liquid helium. He found that it became extremely large at a temperature close to 2.2 K, suggesting a transition to a new liquid state, dubbed liquid helium-II. A graph of the specific heat versus temperature looked like the Greek letter lambda, and so this was dubbed the lambda transition. The actual nature of this state of matter was unknown for a decade. In 1937, two experimental groups independently discovered that helium-II was a superfluid, a fluid that can flow freely without any resistance. One group was John Allen and Don Misener. The second group was led by Kapitsa.

Water flows much faster than honey, which is said to be more viscous. Viscosity is the physical property of a fluid that is a measure of its resistance to flow. We are most familiar with helium as a gas that is used in party balloons. Helium gas has a smaller density than air and so tends to rise. As Kamerlingh Onnes discovered, if helium gas at atmospheric pressure is cooled down to 4 K it becomes liquid. If the pressure is lowered further, the temperature drops further, and the liquid continues to boil. Something quite unexpected happens at temperatures below 2.17 K. The bubbling associated with boiling suddenly stops and the liquid becomes completely calm. Further experiments reveal even stranger properties. The liquid can climb up the walls of a container, a behaviour known as helium creep (Figure 5). In the new state fluid flows without friction (viscosity). For temperatures greater than 2.17 K, liquid helium cannot pass through a filter composed of finely ground powder. However, below 2.17 K, the fluid can pass without any resistance through the filter. The liquid helium

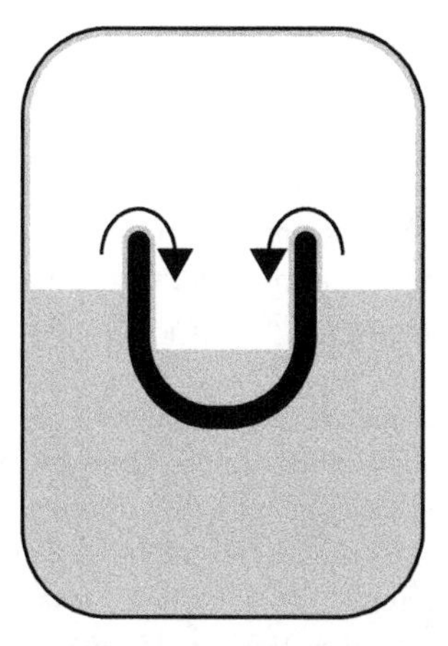

5. Superfluidity. When a U-shaped container is placed in a bath of superfluid helium, the liquid will climb up the outside walls of the container and fill it.

is said to be a superfluid, a distinct state of matter. Kapitsa desperately wanted Landau out of prison so he could explain this phenomenon.

There are two types, or isotopes, of helium atoms in the universe. The most common one is denoted ^{4}He and has an atomic nucleus consisting of two protons and two neutrons. In contrast, ^{3}He has only one neutron, and is extremely rare, mostly being found inside stars. One scientific benefit of the development of nuclear weapons was that it became possible to produce relatively large quantities of ^{3}He gas on earth. Scientists then cooled the gas down until it became liquid. Below temperatures of about 1 K liquid ^{3}He has different properties from liquid ^{4}He. In particular, collectively the ^{3}He atoms move around and act in a manner that has many similarities to how the electrons in a metal behave. Landau had this important insight. This difference between the physical properties of liquids of the two isotopes (^{3}He and ^{4}He) is due to profound reasons, arising from quantum theory. Fundamental particles, such as electrons, protons, and neutrons, can only be of two types, known as bosons and fermions. Electrons and ^{3}He atoms are fermions, which means that two of them cannot be in the same quantum state. In contrast, any number of bosons, such as ^{4}He, can be in the same quantum state. If the total number of protons and neutrons in the nucleus of an atom is an even (odd) number, then the atom is a boson (fermion). Hence, ^{4}He atoms are bosons, while ^{3}He atoms are fermions, respectively.

Liquid ^{3}He does become a superfluid, but only at temperatures below a few milliKelvin. Furthermore, there are three distinct superfluid states, and they all have richer properties than superfluid ^{4}He. These states of ^{3}He can be viewed as combinations of a superfluid, a ferromagnet, and a liquid crystal. A phase transition of liquid ^{3}He was discovered in 1972 at a temperature of a few milliKelvin, using a method like the cooling experiment represented in Figure 4. It was thought at first that the new state was a magnetic solid state not a superfluid state. Regardless of this incorrect identification, Doug Osheroff, the Ph.D. student who made the discovery, was still awarded a Nobel Prize.

In 1995, gases of very low density were cooled down to low temperatures of a few nanoKelvin (one-thousandth of one-millionth of a degree). At these mind bogglingly low temperatures the gases were discovered to be a superfluid, known as a Bose–Einstein Condensate, that occurs in a gas of bosons. At the other end of the temperature scale, the interior of neutron stars, in which matter is extremely dense, becomes superfluid when they cool below about one billion degrees Kelvin.

Later we will see that some properties of superfluids and superconductors manifest the weirdness of quantum theory. The analogues between liquid ^{3}He and metals, and between superfluids and superconductors, are rather surprising because metallic and superconducting behaviours mostly occur in solids not liquids. This is an example of the universality associated with emergent phenomena (materials that are very different at the atomic scale can be in the same state of matter) and partly why solid state physics was renamed condensed matter physics. It is not just about solids, but liquids as well.

Tourist maps for states of matter

Our everyday experience occurs at temperatures around 20 °C and air pressures of about one atmosphere (14.7 psi in terms of the units of pressure often used for inflating car tyres). We make statements such as the following. Copper is a metal and a solid. Mercury is a metal and liquid. Hydrogen is a gas. Carbon dioxide is a gas. Iron is a solid, a metal, and magnetic. However, when the temperature or pressure applied to a material is changed it can enter a different state of matter. Mercury can become a crystal. Helium can become liquid.

If you want to find your way around a city or through a forest a map is helpful. It tells you what you will find where and how different things are located relative to one another. *Phase diagrams* are like maps for states of matter: a picture (two-dimensional plot) that indicates which state is stable at different temperatures and pressures. They show under what conditions transitions between different states (phases) can occur. An example of a phase diagram is shown in Figure 6. It also provides a convenient way to represent information about how the melting and boiling temperatures vary with external pressure. Although chemists can make thousands of distinct compounds, many of their phase diagrams have a similar structure to that shown in Figure 6. For example, the diagrams for water, carbon dioxide, nitrogen, and ammonia are basically the same. The only thing that changes from one material to another is the values of the temperatures and pressures at which specific features in the diagram (such as the critical point) appear and the slopes of the lines.

I should explain what I mean by a state of matter being 'stable' at a specific temperature and pressure. Take some ice out of your

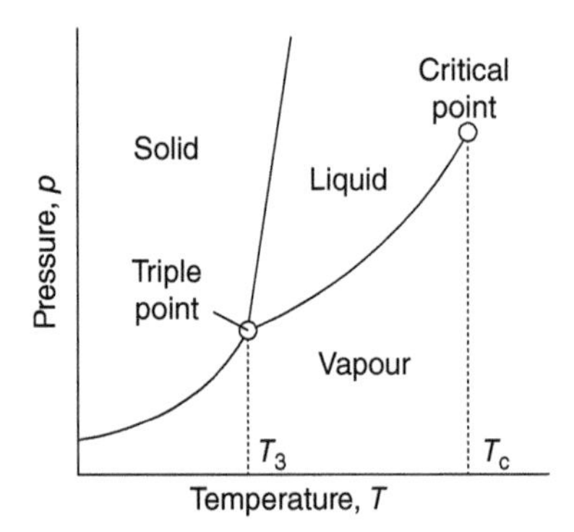

6. Typical phase diagram of a pure material. It shows under what conditions (temperatures and pressures) the different states (solid, liquid, gas) are stable. Along the lines two states can coexist. The line between the solid and liquid states shows how the melting temperature varies with pressure.

freezer and place it in a glass in your warm kitchen. The ice will eventually melt and reach the same temperature as your kitchen. The water is then said to have come to thermal equilibrium with its environment (the air in the kitchen). After that nothing changes, that is, the state of the water is stable. Now consider steam that rises from a boiling saucepan of water and hits a surface in the kitchen. The steam will condense, become liquid water. At room temperature and pressure, if you wait long enough any sample of water will eventually become liquid. Under these conditions liquid is always the stable state of water. On the other hand, inside your freezer, if you wait long enough any liquid water will become ice. Below 0 °C the stable state of water is solid.

The information encoded in phase diagrams reflects an important fact about the material world. If you take a specific substance, whether water or carbon, and fix the external conditions such as temperature and pressure, then the substance must always be in the same state and have the same properties. This is regardless of where in the universe the material is located, how large the sample of material is, or what the previous history of the sample is. All the physical properties are determined: density, compressibility, electrical conductivity, heat capacity, and so on. This universality is of practical significance. A scientist can measure these properties a single time and tabulate them for future reference. Engineers make use of such tables of experimental data for designing a host of industrial processes such as refrigeration, power generation from steam turbines, engines, and distillation.

On the phase diagram shown in Figure 6 the point at the end of the line denoting the boundary between the liquid and gas states is denoted the *critical point*. Above this point there is no qualitative difference between liquid and gas states. On many phase diagrams there is such a point at the end of the line between two distinct states of matter. Critical points are so interesting and important for the conceptual development of condensed matter physics that they will be the subject of a whole chapter.

Something sublime

Figure 6 shows that there is a single temperature and pressure for which solid, liquid, and gas can coexist. This point on the phase diagram is known as the *triple point.* For pure water this occurs at a pressure of 0.006 atmospheres and a temperature of 0.01 °C. Below this pressure a solid does not melt, but rather undergoes a direct transition from the solid state to the gas state. This is known as *sublimation.*

Sublimation is the physics behind 'dry ice', which is actually solid carbon dioxide. When warmed above –78.5 °C the 'ice' is not 'wet' (liquid) because it transforms directly to gas. Sublimation of carbon dioxide occurs at normal atmospheric pressure because the triple point of carbon dioxide is larger than atmospheric pressure.

Sublimation is used in the preparation of freeze-dried food. Sublimation can be used to remove the water from food to preserve it for later use. The food is first frozen at atmospheric pressure. This traps the water in the food as ice. This frozen food is then placed in a 'vacuum', really a container with air pressure less than that of the pressure of the triple point of water. The food is then heated, and the solid water trapped in the food sublimates, escaping from the food as vapour. Thus, the food is dried. The food is then taken out of the vacuum. When someone wants to eat the food, they place it in liquid water and the freeze-dried food absorbs water and returns to its original state.

Phase diagram of carbon

Just like water and carbon dioxide, pure carbon can be found in solid, liquid, and gas states. However, as discussed in Chapter 1, carbon has two distinct solid states: diamond and graphite. The phase diagram of pure carbon (Figure 7) shows that diamond is only stable at very high pressures, about 15,000 times atmospheric

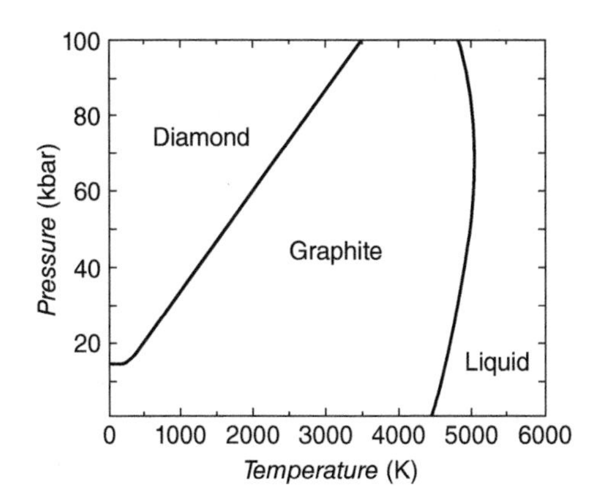

7. Phase diagram of pure carbon. It has two distinct solid states, graphite and diamond. The gas phase occurs at higher temperatures than shown on this figure. The vertical scale is pressure in units of 1,000 atmospheres (kbar) and the horizontal scale is temperature in absolute temperature (Kelvin).

pressure. This seems to go against our everyday experience but the reason is that diamond is 'meta-stable' at atmospheric pressure. If we wait long enough it will convert to graphite. However, that 'long enough' is actually billions of years.

The phase diagram of carbon can help answer several interesting questions. Under what conditions can we make artificial diamonds? Under what conditions were natural diamonds formed? Where might miners look for natural diamonds?

The phase diagram shows that graphite (which is cheap and abundant) can be converted into diamond (which is valuable and rare) by applying pressures greater than 15,000 times atmospheric pressure. This is the basis for making artificial diamonds in a laboratory. But it turns out to be hard, expensive, and slow as the pressures required are so large. Indeed, until recently such high pressures were at the limits of what could be created in a laboratory.

Natural diamonds are formed deep underground, more than 150 kilometres below the surface of the earth, where the pressures are large enough due to the weight of all the rocks above. Fortunately, miners do not have to dig this deep to find diamonds, as the deepest mine in the world is only four kilometres deep. Due to volcanic eruptions some diamonds have been pushed much closer to the surface of the earth.

There are many more phase diagrams I could describe: different kinds of magnets, mixtures of different fluids such as alcohol and water, the quark–gluon plasma in the early universe, and different forms of liquid crystals. Chemical engineers and materials engineers spend a lot of time constructing, studying, and using phase diagrams. This is because they are useful for a wide range of technological applications such as refrigeration, distillation, and production of alloys like steel (a solid mixture of iron and carbon). Fluids in the supercritical state—the state that occurs at pressures and temperatures above the critical point—are used to decaffeinate coffee and to reduce carbon dioxide emissions from coal-powered electricity plants.

This chapter has introduced several key ideas. Different states of matter have qualitatively different physical properties. Even for materials that are composed of just one or two types of atoms they can form many different states of matter. It is not just solid, liquid, and gas. Transitions between distinct states are defined by abrupt changes in properties. Phase diagrams encode which state of a material is stable under specific external conditions such as temperature and pressure.

Just like zoologists and botanists, condensed matter physicists classify things. In the next chapter we consider the concept of symmetry as it is key to characterizing the qualitative difference between different states of matter.

Chapter 3
Symmetry matters

Snowflakes form incredibly diverse structures, seen when they condense onto a plate of glass. Every snowflake is different. On the other hand, every snowflake is the same. They are all composed of ice, a solid state of water. Every snowflake is composed of units that have a sixfold symmetry (Figure 8). Every snowflake is composed solely of water molecules. This paradox of the particular and the universal is at the heart of condensed matter physics. Although diversity prevails anything is not possible. No snowflake has fivefold symmetry. Snowflakes have enchanted scientists for a long time. The astronomer Johannes Kepler studied them and in 1611 wrote a small book about them as a gift for his patron. Kepler suggested snowflakes provided clues to deeper questions about the composition of matter. Today, Kenneth Libbrecht, a physicist at Caltech, has spent most of his career studying snowflakes and has produced beautiful volumes of photographs of them.

Condensed matter physicists ask several questions about snowflakes. What is the reason for the sixfold symmetry of the snowflake? What is the connection between the macroscopic properties of snowflakes and the properties of the underlying microscopic constituents, molecules of H_2O? How is the diversity of snowflake shapes possible? Is there a phase diagram that defines the external conditions under which the different shapes form?

8. A snowflake shows a sixfold symmetry, just like a hexagon. The snowflake appears identical when it is rotated by an angle of 60 degrees about an axis passing through its centre and perpendicular to the page.

There is a long history in art, architecture, philosophy, and science, of associating symmetry with beauty and perfection. The ancient Greek philosopher Plato was a proponent of this view. He studied a particular class of solid shapes: cube, tetrahedron, octahedron, icosahedron, and dodecahedron. Plato identified the first four shapes with the four 'elements': earth, wind, fire, and water, respectively, and the fifth with the heavens. Each of these solid shapes is highly symmetric. Every face of a Platonic solid is the same shape (square, triangle, pentagon…) and each of those shapes has edges of equal length.

Like Plato, Kepler believed that 'God is a geometer' and that God's creation should reflect the perfection of God. These convictions led Kepler to propose in 1597 that the orbits of the planets around the Sun were circular and that the Platonic solids determined the relative size of the orbits. Later this model for the solar system was shown not to be true. In fact, Kepler himself became famous because he showed that the planets moved in elliptical, not circular orbits. Nevertheless, Kepler's model was the beginning of a long history of successfully relating physical laws to symmetry and geometry.

A key discovery in physics from the past century is that symmetry is central to understanding a wide range of physical phenomena, whether colliding billiard balls, the allowed energies of an atom, the fundamental forces of nature, or different states of matter. Symmetries determine what is physically possible. For example, that energy cannot be created or destroyed is a consequence of the fact that physical laws do not change with time.

In this chapter I explore three key ideas. First, transitions between different states of matter are associated with changes in symmetry. Thus, symmetry provides a criterion for specifying the qualitative difference between distinct states of matter. Second, for a specific state of matter the relevant symmetry constrains what is physically possible. Third, symmetry is central to making connections between the macroscopic and microscopic properties of a state of matter. The next chapter will explore how symmetry is associated with the type of ordering that occurs in a state of matter.

The symmetry of an object is defined by the set of transformations (such as rotations and reflections) that leave the object looking just the same. Figure 9 shows specific objects that do not change when rotated by particular angles or when reflected in a mirror. These are examples of discrete symmetries. The objects only appear the same when rotated by specific angles. In contrast, a circle appears the same when rotated by any angle, and so is said to have a continuous rotational symmetry.

Crystals

Naturally occurring minerals are often found as crystals and have fascinated collectors for centuries. Crystals such as diamond,

9. Examples of discrete symmetries. The objects at the top appear the same if they are rotated by specific discrete angles. For example, the playing card looks identical if rotated by 180 degrees. The objects on the bottom are unchanged if they are reflected through a vertical line that passes through the centre of the object.

sapphire, ruby, and amethyst are used in jewellery. The best crystal specimens have clean flat surfaces, like the quartz crystals shown in Figure 10.

Crystallography, the science of crystals, has a long history. In 1669, based on his studies of quartz crystals, Nicolas Steno proposed that 'the angles between corresponding faces on crystals are the same for all specimens of the same mineral'. That would indicate that samples of different minerals might appear different but that on a smaller scale they all had the same underlying structure. One hundred years later, René Just Haüy accidentally smashed crystals of the mineral calcite and observed that the fragments always had the same form as the original crystal. This suggested that crystals are composed of some type of repeat unit, such as a cube or

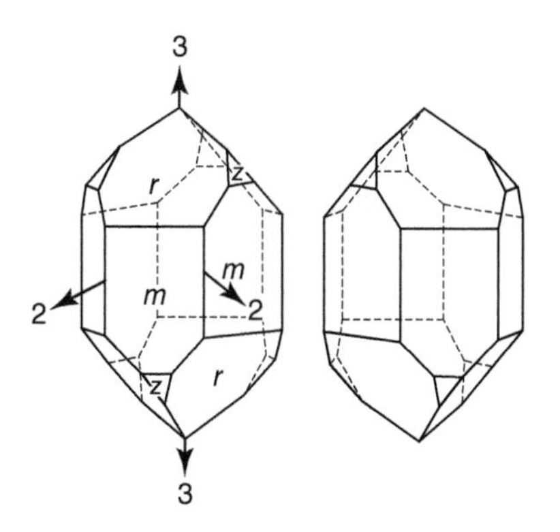

10. Crystals of quartz. Note the flat surfaces. Quartz crystals are found in two different forms, known as left- and right-handed. The two forms are the mirror images of each other.

hexagonal prism. In other words, crystals involve a repeating pattern. If a crystal is infinitely large, then moving (translating) it by a distance the size of the repeat unit leaves the crystal looking the same. Hence, translating the infinite crystal by any integer multiple of the size of the repeat unit leaves the crystal looking the same. This is known as a discrete translational symmetry.

Symmetry limits the number of distinct ways that repeat units can be arranged in a crystal. The mathematics of repeating units was worked out by August Bravais and Arthur Schoenflies in the 19th century. Bravais considered infinite sets of discrete points with translational symmetry, now known as Bravais lattices. Bravais and Schoenflies showed that if translational symmetries are combined with other discrete symmetries such as rotations and reflections, there are only a limited number of possible repeat structures. For example, in two dimensions there are only five possible Bravais lattices: those shown in Figure 11. A lattice with fivefold symmetry is not possible. This is related to the fact that one cannot cover the plane with pentagons without leaving spaces.

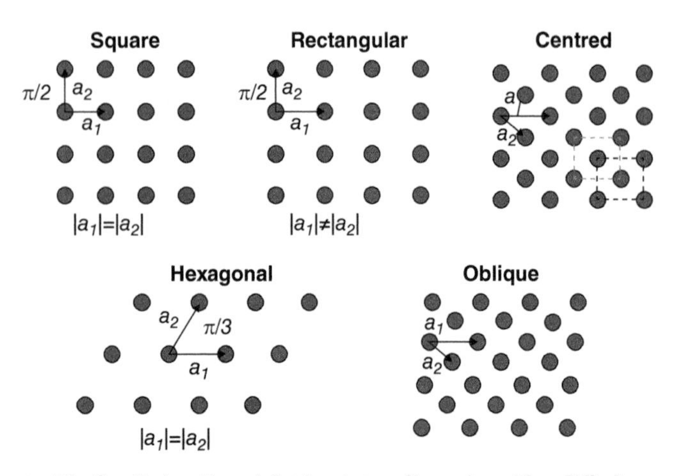

11. The five distinct Bravais lattices in two dimensions. They differ by the amount of rotational symmetry. The rectangular, square, and hexagonal lattices have two-, four-, and sixfold rotational symmetry, respectively. The centred lattice has twofold symmetry, and the oblique lattice has no rotational symmetry.

In three dimensions, there are 14 possible Bravais lattices. Symmetry constrains what is allowed.

At the beginning of the 20th century, a fundamental question was: what are the repeat units in a crystal? The hypothesis that everything is made of atoms was largely accepted, particularly by chemists, and so it was considered reasonable that crystals were periodic arrays of atoms. However, at that time there was no definitive evidence for the actual existence of atoms. Some scientists such as Ernst Mach considered atoms a convenient fiction. After Einstein's theory of Brownian motion in 1905 and experiments of Jean Perrin in 1908, most scientists accepted the reality of atoms. But Mach still held out. The year 1912 saw the development of a revolutionary new method to actually 'see' the atoms inside a crystal. After that, there was no doubt that crystals were periodic arrays of atoms.

X-ray crystallography

In 1912 Max von Laue shone a beam of X-rays on a crystal of zinc sulphide and observed discrete spots on a photographic plate placed behind the crystal. This was taken to be evidence that the X-rays were waves rather than particles, and were diffracted off the atoms in the crystal. X-rays are waves of electromagnetic radiation, just like light and microwaves. Waves are diffracted by a crystal, the process where a wave is scattered by each of the atoms in the crystal. These scattered waves interact with one another to produce a wave that leaves the spots on the photographic plate. These spots arise because of constructive (destructive) interference whereby the X-ray intensity at a particular point on the photo increases (decreases).

The relative arrangement of the atoms in a crystal can be determined from X-ray diffraction for the following reasons. There are only a finite number of distinct crystal structures allowed by symmetry. Each of the 14 allowed Bravais lattices is associated with a unique X-ray diffraction pattern. Thus, from the diffraction pattern one can determine the Bravais lattice. A single atom of a particular chemical element diffracts X-rays in a unique way, for instance, zinc and sulphur atoms scatter X-rays by different amounts. Scientists then work backwards from the X-ray diffraction pattern to the crystal structure.

At the age of 22, Lawrence Bragg developed the relevant mathematics to work backwards from Laue's X-ray diffraction pattern to determine the size, geometry, and symmetry of the repeat unit for crystals of zinc sulphide. He then worked with his father, William Henry Bragg, to determine the crystal structures of sodium chloride (common salt) and diamond.

In *The Cambridge Guide to the Material World*, Rodney Cotterill claims that Laue's X-ray diffraction experiment 'can be regarded as the most important ever undertaken in the study of condensed matter'. When I first read this statement, I thought it was overblown; there are so many other important experiments. But I now agree with him. Knowing the detailed atomic and crystal structure of a material is the first step to address the central question of condensed matter physics, 'How do macroscopic properties emerge from microscopic (atomic scale) properties?' And X-ray crystallography has become crucial in many other fields of science, including mineralogy, chemistry, and structural biology. Most of what is known about molecular biology is based on the structures of DNA and proteins that have been determined by X-ray crystallography.

The crystal structure of the most common form of ice (solid water at one atmosphere pressure) was proposed in 1935 by the chemist Linus Pauling and later confirmed by X-ray diffraction experiments. The structure is shown in Figure 12. The repeat unit is hexagonal, which is why snowflakes have sixfold symmetry. Snowflakes are composed of crystals of water. Each one of these building blocks has sixfold symmetry and so does the snowflake. Each of the 18 distinct

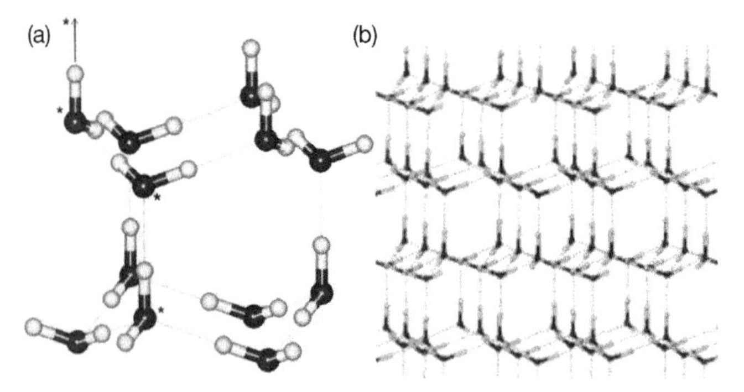

12. The crystal structure of the common form of ice. (a) The repeat unit with the spatial arrangement of water molecules. Black and white spheres represent oxygen and hydrogen atoms, respectively. (b) How these units repeat in the crystal structure to produce a hexagonal structure.

solid states of water has a distinct spatial arrangement of the water molecules and a corresponding distinct symmetry.

The relative arrangement of the carbon atoms in the crystal structures of graphite and diamond is shown in Figure 2. Note that the two structures have different symmetries. In graphite the individual layers have a hexagonal symmetry. In contrast, diamond is not layered and has the same translation symmetries in all three directions. The symmetry around each carbon atom is that of a tetrahedron. Thus, a transition between graphite and diamond, which occurs with increasing pressure, is associated with a change of symmetry.

Quasicrystals

Given that X-ray crystallography is now more than 100 years old, one might expect the field to be a bit boring and predictable. However, condensed matter physics is full of surprises. An example is an unexpected discovery in the early 1980s. Dan Shechtman was studying alloys of aluminium and manganese and performing diffraction experiments to determine their structure. He observed a very sharp diffraction pattern, shown in Figure 13,

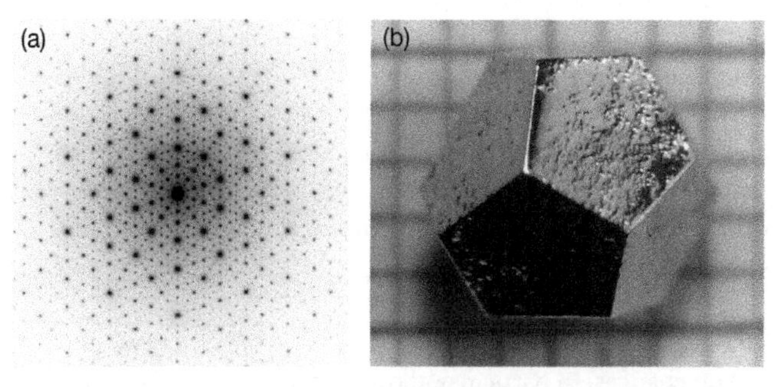

13. Distinct properties of a quasicrystal. (a) Diffraction pattern with fivefold symmetry. (b) Faces that are pentagons. Both of these are impossible for a crystal that is composed of a periodic arrangement of atoms.

with a surprising property: fivefold symmetry. As noted earlier, this is forbidden for a crystal. None of the Bravais lattices shown in Figure 11 have fivefold symmetry. The observed diffraction pattern was so unexpected that it took Shechtman several years to get his results published in a scientific journal. Later, it was shown that a structure that is ordered in a way that is almost periodic, known as quasiperiodic, can lead to such a sharp diffraction pattern. After Shechtman's discovery it was realized that there was a connection to patterns that were originally investigated for fun by the mathematical physicist Roger Penrose in the 1970s. An example of a Penrose tiling is shown in Figure 14.

Shechtman had discovered a new state of matter, a quasicrystal. In 1992 the International Union of Crystallography revised their definition of a crystal. A crystal is no longer defined as a periodic arrangement of atoms, but rather a material that gives a sharp diffraction pattern. In 2011 the Nobel Prize in Chemistry was awarded to Shechtman.

Continuous symmetries

If a square is rotated by an angle of 90, 180, or 270 degrees it appears the same. It has a discrete fourfold symmetry. In contrast, a circle can be rotated by any angle between zero and 360 degrees, a continuous range of values, and it will look the same. A liquid or gas has continuous rotational symmetry. If an observer is inside a huge volume of the fluid, it looks the same in every direction. A fluid also has continuous translational symmetry. It does not matter where an observer is placed inside the fluid, it looks the same at every point.

When a liquid freezes and becomes a crystal the symmetry of the system changes. In particular, the continuous translational and rotational symmetries of the fluid are reduced to the discrete translational and rotational symmetries of the crystal. A liquid

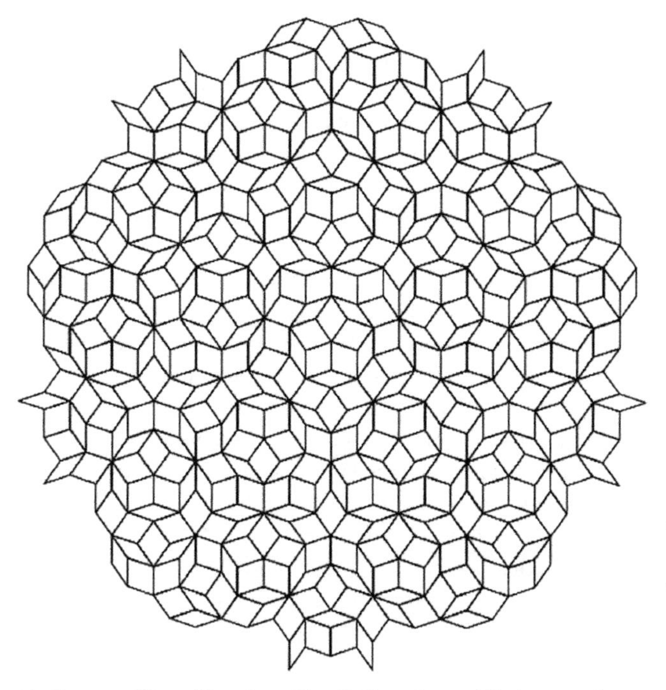

14. Penrose tiling of the plane. Two distinct shapes (tiles) are used to completely cover the plane without any gaps. Note that there are small regions with fivefold rotational symmetry, but the whole tiling does not have fivefold symmetry. The tiling does not quite have translational symmetry. It is not periodic but quasiperiodic.

crystal is associated with discrete rotational symmetries and continuous translational symmetries.

Macroscopic properties reflect microscopic symmetries

The changes in symmetry at the microscopic level that are associated with a distinct state of matter have consequences at the macroscopic level. Symmetry not only constrains what crystal structures are possible, but also constrains the properties of the crystals at the macroscopic scale.

In a fluid (gas or liquid) there are two ways to distort a cubic volume of the fluid. One way is to compress the cube into a shape (a rectangular prism) where the lengths of the sides are not identical, but the angles between the sides of the shape are still 90 degrees. A second distortion changes the shape of the cube to that of a rhombohedron; the angles are no longer 90 degrees, but the lengths of the sides remain the same. Associated with these two types of distortions, there are two distinct ways in which sound can travel through a fluid. Sound waves in air consist of the first type of compression—oscillations in the density and pressure of the air occur in the same direction that the sound wave travels. A second type of sound wave corresponds to the second type of distortion, and is called a shear wave. The two types of sound travel at different speeds. An earthquake produces both types of waves: pressure waves and shear waves, the latter travelling slower. Comparing the two types of waves plays an important role in seismology and in ultrasound imaging in medicine.

The different types of sound that can travel through a particular crystal are macroscopic manifestations of the specific translational and rotational symmetries that the crystal has at the microscopic level. The highest symmetry crystal is a cubic crystal. It has three distinct types of sound. In a crystal with no rotational or mirror symmetry there are many more different sound types. Hence, if one measures all the different types of sound in a crystal, one can gain information about what the symmetry of the crystal is.

Different states of matter are associated with different symmetries. This provides a means to codify the qualitative differences between different states of matter. The relevant symmetry constrains what is physically possible. Symmetry also reflects the connection between the macroscopic and microscopic properties of a state of matter.

The next chapter discusses how the change in symmetry between different states of matter can be quantified and how it is related to the amount and type of 'order' in the specific states.

Chapter 4
The order of things

Life and the world around us sometimes appears chaotic and random. We may feel this way about traffic, weather, economics, social change, politics, or our personal relationships. Perhaps that is why many yearn for regularity, predictability, order, and stability. Science is a search for patterns and order in the natural world. Condensed matter physics is about how order emerges from disorder.

This chapter explores how different states of matter are associated with different types of ordering of the atoms in the material. The symmetry of the state reflects the type of ordering, that is, the patterns associated with the state. There is also a rigidity associated with the ordering and the rigidity determines the nature of the deviations from perfect ordering and results in entities such as vortices that are central to the physical properties of the state of matter.

The association of a state of matter with a specific type of ordering is illustrated in Figure 15 by an analogy involving dodgem bumper cars at an amusement park. A quiet day at the park is not much fun as collisions between cars are rare. In other words, there is little correlation between the relative locations and speeds of the cars. In comparison, on a busy day at the park the spatial

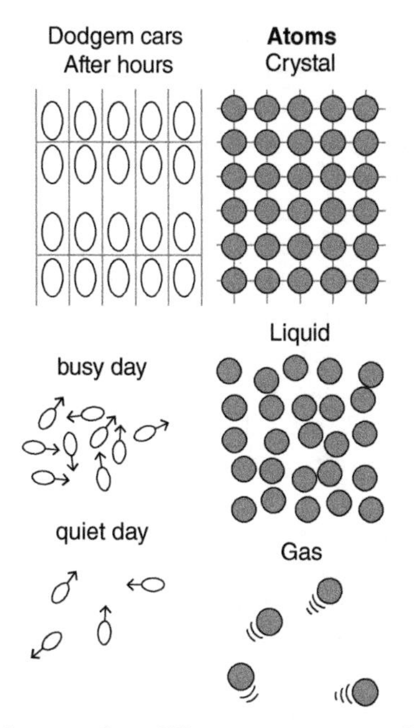

15. An analogy between three different arrangements of dodgem cars at an amusement park and three different states of matter: crystal, liquid, and gas. The only ordered arrangement is for the crystal (park after hours) and this is associated with a specific symmetry and rigidity. The liquid and gas (busy and quiet day) only differ in density and the amount of correlation between the positions of the different atoms (dodgem cars).

separation of the cars is small, and their positions and speeds are more correlated with one another than on a quiet day. But, in both cases, there is no ordered arrangement of the cars. In contrast, after the park closes the cars are parked and arranged in an orderly manner. There is a rigidity associated with their spatial arrangement. One car cannot be moved without moving others. These three states of the dodgem cars are analogues of three states of matter: gas, liquid, and crystal.

In the analogy using dodgem cars, there are other possible types of ordering. In some amusement parks there is a track, and the cars are meant to all go in the same direction. The symmetry between clockwise and anti-clockwise of the track is then broken. In the car park, Figure 15 shows cars that are symmetrical with respect to front and back. However, real cars have a front and back, and so can be parked either front first or back first. Hence, several types of ordering are possible: all cars park back first, all cars park front first, cars are front first or back first at random, alternating patterns of front first and back first as one goes along a row, alternating rows of front first and back first, and so on. These different types of ordering in the car park all have analogues in different solid states of matter.

Liquid crystals involve unique types of ordering. These materials are composed of elongated organic molecules, such as those shown in Figure 16. At high temperatures the material is in a liquid state and the orientations and positions of the molecules are random. The liquid has both continuous translational and rotational symmetry. At low temperatures the molecules form a solid crystal without the continuous translational and rotational symmetry of the liquid state. As the crystal is heated the temperature increases and there is a phase transition to the liquid crystal state, in which all the molecules point in the same direction, but their positions are random. Hence, the liquid crystal state has the continuous translational symmetry of the liquid, but not its continuous rotational symmetry, like the crystal. As the temperature increases further there is a transition to the liquid state (Figure 16). In terms of the analogy using dodgem cars the liquid crystal state is similar to when cars park in a field all pointing in the same direction but there are no grid lines, and their positions are then random.

The existence of a state in between a liquid and crystal was first proposed in 1888 by botanist and chemist Friedrich Reinitzer who

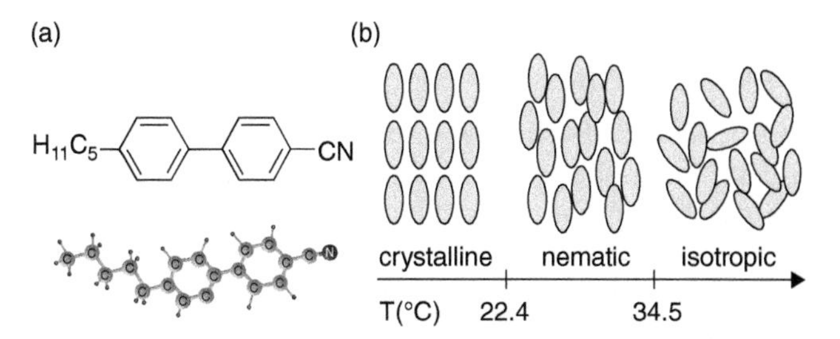

16. Liquid crystals. (a) An example of the type of elongate organic molecule found in these materials. Each molecule can be represented by an oval shape. (b) In the nematic liquid crystal state, the molecules tend to point in the same direction, but their positions are random.

was doing research on cholesterol at the Institute for Plant Physiology in Prague. He performed a heating experiment similar to that described in Figure 4. Instead of one melting transition he observed transitions at two distinct temperatures.

There are multiple alternative orderings for liquid crystals with names such as nematic, smectic, chiral nematic, discotic, and chlorestic. In the smectic phase molecules form layers of oriented molecules. The character of the liquid crystal state can be detected by shining polarized light on the material. Liquid crystal displays (LCDs) in electronic devices use the property that an electric field can orient the molecules, and this changes the interaction of the material with polarized light.

For solid crystals the nature of the ordering and the symmetry associated with a specific crystal structure is clear once the spatial arrangements of the atoms in the crystal are determined, such as by X-ray diffraction. For other states of matter, such as superconductors, superfluids, and antiferromagnets, the nature of the ordering and the symmetry is often not apparent and has only been determined with significant scientific insight.

Magnetic order

The type of magnetism that occurs in solid iron is known as ferromagnetism. The prefix ferro is derived from the Latin word *ferrum* for iron. To understand the nature of ferromagnetism and the associated order and symmetry it is worth recalling the basic nature of magnets, familiar from school science classes (Figure 17). Iron magnets are associated with a definite orientation; they have a 'north pole' and a 'south pole', and when two magnets are placed close to one another, unlike poles attract and like poles repel one another. This is why a compass needle that is placed near a magnet will settle at a direction parallel to the direction of the magnetic field produced by the magnet. A steel needle or paperclip contains iron but is not magnetic unless it is placed in contact with a magnet. The steel needle is composed of many spatial regions called domains; in each domain the magnetism is oriented in a specific direction, but this direction is different in different domains and their

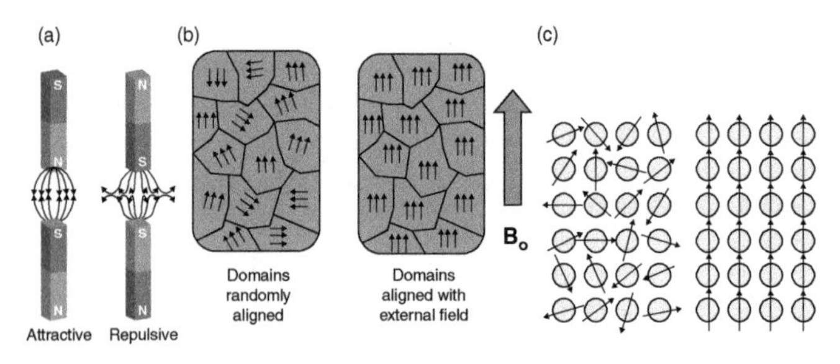

17. Ferromagnetic order. (a) In a common magnet, ferromagnetism is associated with a particular direction, denoted by a north and a south pole. Opposite (like) poles of two magnets attract (repel) one another. (b) A piece of a ferromagnetic material may contain many different domains, each of which has a particular magnetic direction. If these domains are randomly oriented, they can be aligned in the presence of a large external magnetic field. (c) In a non-magnetic state the atomic magnetic dipoles are oriented in random directions (left). The order in the ferromagnetic state is that the dipoles are oriented in the same direction (right).

magnetic effects tend to cancel one another out. However, rubbing the needle on the magnet aligns all these directions so that the needle becomes magnetic. Each domain is also composed of large numbers of sub-regions, each with a particular magnetic direction.

This subdivision of the magnetism goes down to the level of the individual iron atoms that make up the material. Each atom is actually like a tiny compass needle, with a particular direction and magnitude of magnetism, known as the magnetic dipole. This is due to all electrons having a 'spin' and is described by quantum theory. The magnetism of a single domain is the sum total of the magnetism associated with all of the iron atoms in the domain. Ferromagnetic order occurs when all these atomic 'compass needles' do not point in random directions, but a finite fraction of them point in the same direction. This is similar to how in a car park there may be a majority of cars that are reverse parked.

Magnetic order similar to that in iron is also found in a wide range of materials and is still called ferromagnetism. It is one of many different distinct states of magnetism that have been discovered. They include antiferromagnetism, ferromagnetism, canted antiferromagnetism, and ferrofluids. In a single material transitions between some of these different states can be induced by changing temperature, pressure, or the strength of an external magnetic field. The different magnetic states are associated with different orderings of the directions of the magnetic dipoles relative to one another. For example, in antiferromagnetism the dipoles of neighbouring atoms point in opposite directions. Finding what these relative arrangements are can be achieved by scattering an intense beam of neutrons off a sample of the material. This was not possible until the 1950s when nuclear reactors were used to produce the necessary beams of neutrons. Like the production of large amounts of helium isotopes, this was another scientific benefit of the development of nuclear weapons.

If a block of iron is heated the strength of the magnetism decreases with increasing temperature until at a temperature of 770 °C (1043 K) the magnetism completely disappears. At room temperature the rare earth metal dysprosium is not magnetic. However, if it is cooled down to a temperature of 88 K it becomes magnetic. All ferromagnets have this property: at a particular temperature, called the Curie temperature, there is a transition between magnetic and non-magnetic states. Magnetism can also be induced in these materials by placing them in the presence of an external magnetic field. On a phase diagram of magnetic field versus temperature, the point on the diagram at zero magnetic field and the Curie temperature is a critical point, just like that for the liquid–gas transition. At temperatures larger than the Curie temperature there is no magnetic order as the atomic magnets all point in random directions.

Back to the USSR

Lev Landau was introduced in Chapter 1 as the first condensed matter theorist with papers that he published in 1937. Landau made notable contributions in all areas of theoretical physics, not just condensed matter. There are more than 20 results, equations, or phenomena that bear Landau's name. For Landau's 50th birthday his colleagues presented him with two tablets inscribed with 'Landau's ten commandments' (his 10 most significant mathematical equations). With his former student, Evgeny Lifshitz, Landau co-authored a classic 10-volume series, *Course in Theoretical Physics*, that is still a standard reference. In the USSR, Landau founded a School of Theoretical Physics that produced a plethora of distinguished theoretical physicists. Tragically, Landau's scientific career ended after a car accident when he was 52 years old. He died six years later from associated injuries.

Landau's papers in 1937 were concerned with developing the simplest possible theory that could describe the properties of a material near a critical point in the phase diagram, such as that associated with a liquid–gas transition or a ferromagnet. A key assumption that Landau made was that most of the microscopic details, such as the chemical composition of the material, are not relevant. Landau introduced an *order parameter* to quantify the amount of ordering present and the symmetry of the ordering. The order parameter varies with temperature and other external parameters such as pressure or magnetic field. The order parameter only has a non-zero value in the ordered state. Landau wrote down the simplest form for the energy of the system as a mathematical function of the order parameter. Symmetry significantly constrains the possible forms for this function. The form is qualitatively different for temperatures above and below the critical temperature. From his theory, Landau obtained mathematical expressions for how the order parameter varies with temperature and how there should be an abrupt change in the specific heat at the critical temperature. What was particularly important was the idea of *universality*: that most of the microscopic details did not matter and that a wide range of materials and states of matter should have similar properties. The ideas in this paper of Landau were foundational for the important ideas about the critical point and universality that are discussed in Chapter 6.

A significant application of Landau's approach to ordering and phase transitions came in 1950. Vitaly Ginzburg and Landau proposed a theory that could describe many of the properties of superconductors, including how they behaved in the presence of a magnetic field, in thin films, and how a large electrical current could destroy the superconductivity. Although the Ginzburg–Landau theory could explain a wide range of superconducting phenomena, it left many questions unanswered, including the actual nature and origin of the ordering associated with the superconducting state.

A key concept in Landau's theory is *spontaneous symmetry breaking*. In some systems the most symmetrical state is unstable and so the system will spontaneously relax to a state with less symmetry. This can occur even though the components of the system and the interactions between them have the full symmetry. This can be illustrated by an example given by Frank Close in his book *Lucifer's Legacy: The Meaning of Asymmetry.* Imagine that you are at a dinner party and seated with a group of people around a circular table. On each side of you there is a napkin (a serviette in some countries). All guests and napkins are equally spaced apart. Which napkin will you take? Left or right? The system has left–right symmetry. At some point one dinner guest decides and picks up the napkin on their left, leading to both of their neighbours also picking up the napkin on their left. This preference for left spontaneously moves through the whole system. The original left–right symmetry has been spontaneously broken. On the other hand, it is equally likely that the right preference could have emerged.

A surface like a Mexican hat can be used in an analogy to illustrate spontaneous symmetry breaking (Figure 18). The surface does not change when it is rotated by any angle with respect to a vertical axis that passes through its centre. Thus, the hat has a continuous rotational symmetry. If a ball is placed on the top of the hat, the whole system (ball plus hat) also has this rotational symmetry. However, this state is unstable, and the ball will spontaneously roll down into a stable location in the trough of the hat, as shown by the arrow. The state of the system then breaks the rotational symmetry. Note that the ball could roll in any direction and so there are an infinite number of possible stable states.

Landau's theory can be illustrated using this analogy. The order parameter is represented by the horizontal coordinates of the ball and the energy of the system is represented by the vertical coordinate of the ball. When a system is at a temperature less than the critical temperature the relevant surface is like a Mexican hat. In contrast, when the temperature is greater than the critical temperature the surface is like a dish where the ball stays at the centre.

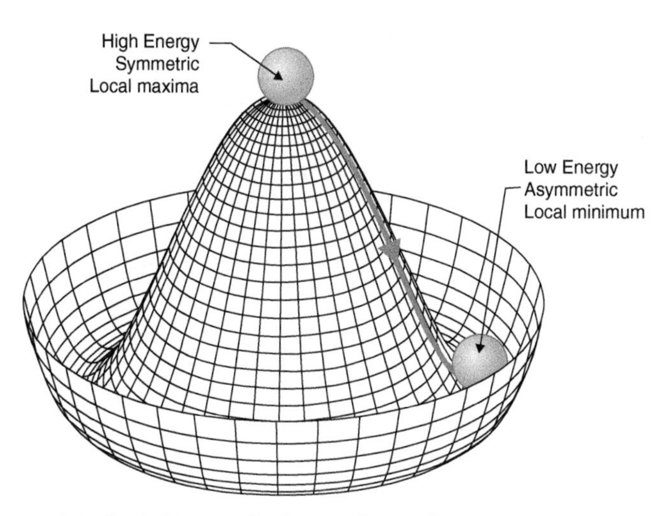

18. A mechanical system that is an analogue of spontaneous symmetry breaking.

Landau's theory is beautiful, both for its simplicity and its unifying power. It provides the paradigm to understand the plethora of states of matter that have been discovered. The value of the order parameter is specified by a finite set of numbers. In the analogy shown in Figure 18, there are two numbers: the horizontal coordinates of the ball. I now briefly discuss some specific examples of order parameters.

For superconductors and superfluid ^{4}He the order parameter is specified by two numbers. For a ferromagnet three numbers are needed to specify the order parameter. These three numbers can be used to represent the direction and length of an arrow and specify the direction of the magnetic dipoles of the atoms in the ferromagnet. There are other types of ferromagnets where the arrow must lie within a plane or along a single direction in the crystal, due to interactions of each atom with the surrounding atoms in the crystal.

In the nematic liquid crystal state, the order is defined by the alignment of the molecules (shown as ovals in Figure 16). However, unlike in ferromagnets, there is no up–down distinction and so the order parameter is not an arrow, but a line with a direction but no arrowhead. This is like the car park in Figure 15 where cars have an orientation but the front and back are identical.

Antiferromagnetism is a state of matter that occurs in crystals of many oxides of transition metals, such as nickel oxide. It is similar to a car park where the cars are in an alternating pattern of front end in and back end in. The magnetic atoms in the crystal can be divided into two different classes, A and B, and each class is associated with a particular subset of sites in the crystal. There are equal numbers of atoms in the two classes. All of the atoms are magnetic and have an arrow associated with their magnetism. At the A sites the arrows point in the opposite direction to the arrows at the B sites. Thus, the whole crystal has no net magnetism, but there is still magnetic order. The order parameter for an antiferromagnet is defined to be the difference between the direction of the arrows on the A sites and of the arrows on the B sites. Louis Neel proposed the existence of the antiferromagnetic state in the 1930s and was awarded the Nobel Prize in Physics in 1970.

There are three distinct superfluid states in ^3He, creatively known as A, A_1, and B. They can be viewed as a combination of superfluid, ferromagnet, and liquid crystal. Eighteen numbers are required to specify the state. The order parameters of the A and B states were deduced by Tony Leggett within a few years of the discovery of superfluidity. In my Ph.D., I investigated how ultrasound could be used to probe the 18 different ways that the order parameter can oscillate.

Rigidity of ordered states

An intuitive idea of what distinguishes a solid from a liquid is that of *rigidity*. A solid keeps its shape, whereas a liquid does not. You have to keep liquids in containers. Rigidity is related to how easy it is to distort a crystal away from a perfect shape, to compress it, stretch it, or to twist it. These distortions change the order and symmetry in the crystal. The distortions move the atoms away from their periodic arrangement.

The concept of rigidity has a natural generalization to other states of matter and is useful for understanding their properties. The generalized rigidity is a measure of how much energy is required to change the value of the order parameter in one part of the system relative to another.

In a material that is in a particular state with an associated order and symmetry, the system will not be perfectly uniform, either spatially or chemically. For example, a crystal of diamond may have defects such as a single carbon atom missing at a particular location, or where there is a phosphorus atom instead of a carbon atom. Such defects are of practical significance. The colour of most minerals used in jewellery, such as ruby or amethyst, is actually not the colour of a perfect crystal of the mineral. Rather the colour is due to defects associated with chemical impurities. The plasticity and brittleness of metals is largely determined by defects known as dislocations, which are associated with disruptions of the translational symmetry of the crystal. Figure 19 shows an example for a two-dimensional crystal where the planes of atoms in the crystal are misaligned.

Topological defects

Topology is the study of the properties of curves and surfaces that do not change when they are deformed by bending or twisting, as opposed to ripping, cutting, or gluing. Smoothly adjusting the position of the atoms can change the location of the defect but cannot remove it. Hence, it is an example of a topological defect. For a three-dimensional crystal, if planes of atoms, such those shown in Figure 19, are stacked on top of one another, the position

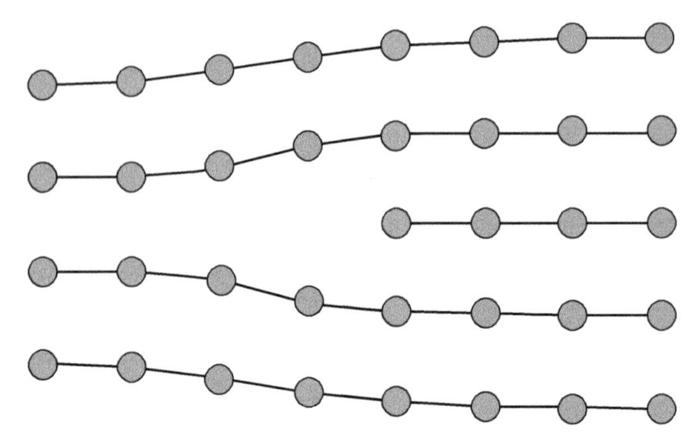

19. A defect in a two-dimensional crystal. The middle row of atoms on the right ends halfway through this region. In the presence of this defect the crystal only has perfect order and translational symmetry a long distance from the defect. Moving atoms around cannot change the fact that four rows of atoms enter from the left and five rows leave to the right.

of the defect defines a line, and the defect is known as a line defect.

A magician bends a thin tube made from copper and gives it to a strongman who is unable to bend it back. This is not magic but science. The original tube is made of soft copper and is relatively free of defects. Bending the tube creates many line defects. They become entangled with one another, just like the strings that make up a rope. Bending the tube back is hard because it requires untangling these line defects. This phenomenon is known as work hardening and is important in metallurgy.

Earlier the dinner napkin problem was used to illustrate the spontaneous breaking of left–right symmetry. Two possible states of the system were considered: one where all the diners used the napkin on their left, another where they all took the napkin on their right. The same system can also illustrate topological defects. Besides the all-left and all-right napkin arrangements, there is another (less optimal) possible napkin arrangement that could emerge when there are a large number of guests. Suppose at the same time that one guest picks up the right napkin a second guest some distance away picks left, and their neighbours follow their lead. This will lead to the table being divided into a 'left' region and a 'right' region. At the boundaries between these regions two defects emerge: one guest will end up with two napkins and another guest with none.

When water goes down a plughole in a bathtub sometimes it forms a whirlpool; in the middle there is no liquid. This structure is also known as a vortex. Vortices also occur as topological defects in superfluids and superconductors. In a superfluid the middle of the vortex, the core, is a normal fluid and the superfluid outside the core is a perpetual whirlpool.

In a particular class of superconductors, known as type II, vortices can form when the superconductor is placed in an external magnetic field. In the core of the vortex there is no superconductivity, the magnetic field penetrates the metal, and there is a perpetual whirlpool of electric current outside the core (Figure 20). The existence of vortices is a natural consequence of the Ginzburg–Landau theory of superconductivity and was first shown in 1953 by Alexei Abrikosov, a colleague of Landau in the Soviet Union.

Here are two interesting anecdotes associated with Abrikosov's discovery. Landau did not believe that vortices could exist, became angry with Abrikosov, and said his work was 'pseudoscience'. A discouraged Abrikosov did not publish his results, showing the esteem in which Landau was held. However, Landau changed his mind in 1955 when Richard Feynman showed that vortices could

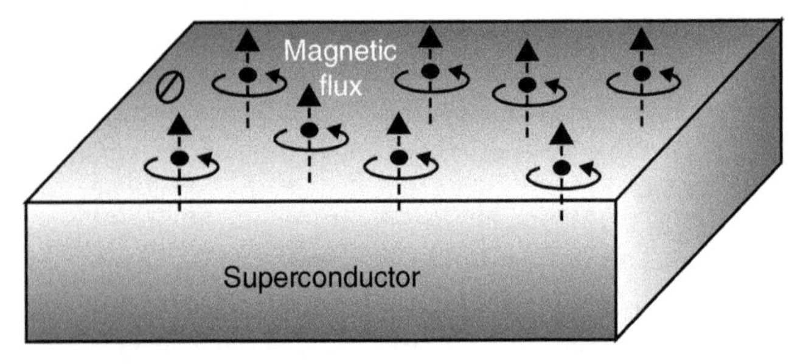

20. Vortices in a superconductor. When a superconductor is placed in an external magnetic field, vortices form. Around the vortex there is a perpetual circulating electrical current and inside the vortex core the magnetic field penetrates the superconductor.

exist in superfluid helium. This led to Abrikosov finally publishing his results in 1957.

Abrikosov's mathematical analysis showed that the number of vortices present in a superconductor would be proportional to the magnitude of the external magnetic field, and that the cores of the vortices would be arranged in a spatially ordered pattern like a square lattice. But Abrikosov made an arithmetic error in his calculations and the most stable spatial pattern for the vortices is actually hexagonal, not square. This hexagonal pattern was first directly observed in an experiment in 1964. Despite these setbacks, Abrikosov shared the Nobel Prize in Physics with Ginzburg and Tony Leggett in 2003.

Given the rich structure of the order parameter in the superfluid phases of liquid ^{3}He, the topological defects that are possible are diverse and complex. For example, the core of vortices can be magnetic even though no external magnetic field is present. Theorists, including David Mermin and Phil Anderson, considered what type of defects would be possible in a droplet of the superfluid A phase. They found that at the surface of the droplet a defect could exist that causes a superflow to slowly decay. Mermin delights in intellectual mischief and named the defect a 'boojum' after a fictional animal that slowly fades away in the nonsense poem *The Hunting of the Snark*, written by Lewis Carroll, best known as the author of *Alice's Adventures in Wonderland.*

Spontaneous symmetry breaking is also important in other sub-fields of physics. It is key to understanding elementary particles (such as electrons, protons, pions, quarks, neutrinos, and so on) and the relationship between different fundamental forces (such as electromagnetism and the weak nuclear force). In 1960, Yoichiro Nambu used some similarities between the theories of superconductivity and of elementary particles to explain why neutrons, protons, and pions have the mass that they do. In the following decade, spontaneous symmetry breaking was central to the development of the 'standard model' for elementary particles and fundamental forces. The final piece of experimental evidence for the standard model was the discovery of the Higgs boson in 2012. This is the particle associated with oscillations in the Mexican hat potential in the direction perpendicular to the circular trough (Figure 18).

Distinct states of matter are associated with distinct symmetries and types of order. Landau's theory quantifies the type and amount of ordering with an order parameter; it is associated with spontaneous symmetry breaking. The rigidity of the state determines the length scales associated with spatial variations in the order parameter. Associated with each kind of order parameter there are new entities, including topological defects.

Everything discussed so far has involved materials that are three dimensional. The next chapter explores the world of two dimensions, dubbed 'Flatland' after the novella about two-dimensional life by Edwin Abbott, and the new states of matter found there.

Chapter 5
Adventures in Flatland

In everyday life we think of most objects as having three dimensions. But what would life be like in a two-dimensional world? For one thing, it would be harder to move around. We could no longer step over things but would have to move around them. In 1884 Edwin Abbott published *Flatland: A Romance of Many Dimensions*, under the pseudonym A. Square, a satirical novella about social life in Victorian England. People are represented by geometrical objects. Men are represented by shapes such as triangles and hexagons. Women are represented by lines. The social status of men increases with the number of sides that their shape has and how many of the sides are of the same length. Abbott's book created limited interest and was largely forgotten by the 1920s. Interest revived when theoretical physicists started to think about worlds in different dimensions. This interest was stimulated by Albert Einstein's theories of relativity, which proposed that we live in a four-dimensional world, not a three-dimensional one. Time is the fourth dimension, and there is an intimate and concrete connection between time and space. Attempts to unify gravity with other fundamental forces has led to physicists proposing and studying theories with more than four dimensions.

Changing the number of spatial dimensions leads to different physics because it changes what is mathematically possible. In

Chapter 3 it was noted that in three dimensions, there were only five highly symmetrical shapes known as Platonic solids (tetrahedron, cube, octahedron, icosahedron, and dodecahedron). In contrast, in two dimensions it is possible to make an infinite number of symmetrical shapes, known as regular polygons, shapes made of straight lines of equal length such as squares or hexagons. Similarly, the number of Bravais lattices differ in two and three dimensions. Changing the number of spatial dimensions changes both what is mathematically possible and what is physically possible.

What would condensed matter physics be like in Flatland? This question received limited attention before the 1970s. Occasionally, theoretical physicists would investigate mathematical models of crystals or magnets in one or two dimensions just because the mathematics was simpler and more tractable than in three dimensions. The goal was to obtain insight into physics in three dimensions. We will consider a famous example, the Ising model.

In the 1970s several surprising developments led to significant interest in condensed matter physics in spatial dimensions different from the usual three. First, it became possible to make a wide range of material systems that were two dimensional. Secondly, theoretical work showed that states of matter, and phase transitions between them, can be qualitatively different in one, two, and three spatial dimensions. And thirdly, considering different numbers of spatial dimensions turned out to be very fruitful for theory, particularly for understanding phase transitions near critical points.

The Ising model

Concrete mathematical models need to be simple enough for mathematical analysis or simulation on a computer. But they need to be complicated enough that they can capture the essential details of phenomena in a real material. This approach has proved key to answering the big question of condensed matter physics, how microscopic and macroscopic properties are related.

The Ising model is a paradigm for this modelling approach. It can be used to illustrate key concepts such as phase transitions, spontaneous symmetry breaking, critical points, ordering, and abrupt changes in physical properties. Although originally proposed to describe magnetic phase transitions, Ising models are now applied to a wide range of topics in physics, chemistry, neuroscience, biology, sociology, and economics.

For a magnetic system each magnetic atom is represented in the Ising model by a single square box that can have two possible states, here represented as black or white, corresponding to two possible directions, 'up' or 'down', for the magnetic dipole of the atom. These two directions can also be viewed as representing whether the magnetic dipole is parallel or anti-parallel to the direction of an external magnetic field. In the model there is an energy gain when two adjoining boxes are in the same state, that is, they are both black or both are white. This energy gain captures a tendency towards ferromagnetism, where the directions of all the dipoles align with each other. In the absence of an external magnetic field, there is equal probability of a box being black or white. The random jiggling associated with the temperature of the system means the state of individual boxes is constantly changing. There is competition between this jiggling and the tendency for adjoining boxes to have the same colour.

The Ising model was originally proposed in 1920 by Wilhelm Lenz, a professor at Hamburg University in Germany. He suggested investigation of the model as a doctoral research project for a student, Ernst Ising. The one-dimensional version of the model consists of a chain of boxes, each of which can be black or white. In his doctoral thesis Ising was able to solve this version of the model exactly. What this means is that he calculated all the properties of the model without making any approximations in his mathematical analysis. He found that even for very low temperatures the model never undergoes a phase transition to an ordered ferromagnetic state. That is, there are always equal numbers of white and black boxes. Ising also gave a rough argument that this would also be true in two and three dimensions. This led to doubts as to whether the model could describe the phase transition between ferromagnetic and non-magnetic states that occurs at a critical temperature in ferromagnets such as iron. This fundamental question was not answered definitively for 15 years, when it was shown that the Ising model does indeed describe such a phase transition in two or more dimensions.

In 1944 Lars Onsager, having performed a mathematical *tour de force*, published an exact solution to the Ising model on a two-dimensional square lattice. He showed that the model did have a critical point at a non-zero temperature and calculated some physical properties of the model. They were different from Landau's theory (Chapter 3), raising questions about the validity of Landau's theory.

A computer simulation of the model on a two-dimensional square lattice was used to produce the pictures in Figure 21. It shows likely states of the system for three different temperatures. The system shown consists of a grid of 124 × 124 small boxes. The probability of a box being black or white depends on the temperature and on whether its nearest neighbours are black or white. At very low temperatures, there is little jiggling, and the most likely state of the system is one where it is all black or all white.

Below the critical temperature large magnetic domains form (Figure 21 (a)). There is more white than black, representing spontaneous symmetry breaking. At the critical temperature there are equal numbers of black and white but there are paths through the whole system that pass through purely white or black

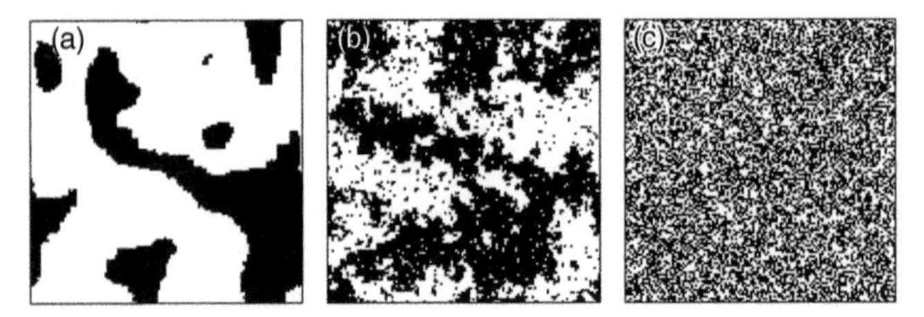

21. Computer simulation of a two-dimensional Ising model. (a), (b), and (c) show a snapshot of a likely configuration of the system at a temperature less than, equal to, and greater than the critical temperature, respectively.

(Figure 21 (b)). This means that the correlation between the colour of two boxes can be strong even when the spatial separation of the two boxes is large. Above the critical temperature, the range of correlations is short (Figure 21 (c)).

On the question of how macroscopic properties of a state of matter emerge from the microscopic properties, Onsager's solution of the Ising model provided a concrete example of an important idea: *short-range interactions can lead to long-range order.* Prior to Onsager's solution, not all experts were convinced that this was true. In the Ising model each square box (magnetic atom) only interacts directly with its nearest neighbours, that is, the interactions are short-range. Yet in spite of all the jiggling the whole system can form a state where changing the state of one atom is correlated with the state of another atom infinitely far away, that is, the system has long-range order. To put it another way, the system has rigidity, similar to the way in which, in a solid object, pushing an atom on one side of the solid will force the atoms on the other side of the object to move. The Ising model is a paradigm of emergence: simple rules can lead to complex properties.

Although Ising is a well-known name in physics he did not go on to a distinguished career in research. After his doctorate, Ising worked in Germany as a high school teacher but lost his job because he was Jewish. He fled to Luxembourg and worked as a shepherd and a railway worker, before emigrating to the USA in 1947. He then worked until retirement teaching physics at a small university and did not resume research.

Onsager was a professor at Yale University and was awarded the Nobel Prize in Chemistry in 1968 for work on irreversible thermal processes. Yet, it is doubtful that Onsager would have survived in today's 'publish or perish' academic culture. He only published one or two papers each year, and some were only a few pages long. But many of Onsager's papers were seminal. He was also slow to publish. Often Onsager would announce his latest research results at a conference, then others would reference them in their own papers, and several years later Onsager would finally publish his results. For example, he announced his solution of the Ising model two years before it was published. Onsager was also known as being difficult to understand, even by brilliant colleagues and graduate students. Before Onsager was hired by Yale, he was fired by both Brown University and Johns Hopkins University because his teaching was so poor.

Creating Flatland in a laboratory

As a result of the painstaking work and ingenuity of many scientists, it is possible to fabricate materials that are essentially one-dimensional or two-dimensional. For example, the electrons in these materials can only move in one or two directions. We have already come across one example of a two-dimensional material: graphene, a single layer of carbon atoms.

Suppose a perfect crystal is cut in two. The surface of a piece is two-dimensional. It can be used as a substrate to create a two-dimensional system. For example, liquid helium placed on a graphite surface can be adjusted in quantity to create a single layer of helium atoms. This has led to studies of superfluids in two dimensions showing that the nature of the superfluid state and the transition to it are different from that in three dimensions.

It is also possible to make layered materials to which layers of atoms are added one layer at a time. By varying the chemical composition of different layers, interfaces can be created with unique two-dimensional properties. A specific example of this is called a semiconductor heterojunction, in which a gas of electrons is trapped at the interface between two different semiconductor materials. The electrons can then only move in directions parallel, but not perpendicular, to the interface. These two-dimensional electron gases exhibit new states of matter, known as quantum Hall states, which we will come to later.

Another class of materials is described as quasi-two-dimensional. They are three-dimensional crystals composed of layers of atoms in which there are only weak interactions between the neighbouring layers. Graphite is an example; it is a crystal composed of layers of graphene. If the interactions between the layers are weak enough, many of the properties of the material may be similar to those of a single layer. Later in this chapter, we will look at a class of superconducting materials that are quasi-two-dimensional. Chemists have also made crystals that consist of chains of atoms and are classified as quasi-one-dimensional materials.

New states of matter

All the atoms in a three-dimensional crystal are jiggling around randomly because of thermal effects. As the temperature increases eventually the jiggling of the atoms becomes so large that the crystal melts (or sublimates) and the atoms can move freely. In spite of all this jiggling, at temperatures lower than the melting temperature the crystal is rigid and on average the atoms are arranged in a periodic fashion. In the crystal there is discrete translational and rotational order. This means that if we know the spatial positions of a few atoms we also know the positions of all the atoms that are a very long distance away. In the liquid, there is neither translational nor rotational order.

The situation in two dimensions is qualitatively different. Due to the thermal jiggling of the atoms, crystals are not stable. It is impossible to have an infinite array of atoms arranged in a perfectly periodic manner. Rather, at low temperatures the system has long-range rotational order but no long-range translational order. This means that unlike in three dimensions an X-ray diffraction experiment will not reveal a series of well-defined spots on a photographic plate. That crystals cannot exist in two dimensions was first shown theoretically in 1968 by David Mermin.

The 'melting' of a two-dimensional 'crystal' is qualitatively different from a three-dimensional crystal. As the temperature increases there is no direct transition from the 'crystal' to a liquid. Rather, there is an intermediate state, known as the hexatic state, that has orientational order but no translational order. The existence of this new state of matter was proposed theoretically in 1979 by Peter Young, David Nelson, and Bert Halperin. As the temperature increases, there is a further phase transition to the liquid state, in which the orientational order is no longer present. Thus, in two dimensions 'melting' occurs in two stages.

The hexatic state was observed experimentally several years later in two different systems. One system was a thin film of a liquid crystal. The second system consisted of latex spheres suspended in water. The spheres had a diameter of about one micrometre and the water was a thin film about one micrometre thick. The fact that the chemical composition of these two systems is very different but they exhibit the same states of matter is an example of the universality common in condensed matter physics. The discovery of the hexatic state is one of the few examples where a new state of matter was predicted to exist by theoretical physicists before it was observed in the laboratory.

The mechanism of melting in two dimensions is qualitatively different from that in three dimensions. In the latter, a crystal melts because at the melting temperature the random spatial motion of the individual atoms becomes as large as the distance between neighbouring atoms and a periodic arrangement of the atoms becomes unstable. In two dimensions, there are topological defects, such as that shown in Figure 19, that are present in pairs. In the low-temperature state these pairs are bound together, just like pairs of positive and negative electrical charges. At the transition to the hexatic state these pairs of dislocations become unbound, and the dislocations proliferate throughout the system, destroying the quasi-long-range translational order.

The main physical ideas described above were first formulated for magnets, superconductors, and superfluids, rather than for solid–liquid transitions. At a finite temperature in two dimensions, it is not possible to have a superconducting, superfluid, or magnetic state with true long-range order. This type of ordering is destroyed by thermal fluctuations. In the early 1970s, Vadim Berezinskii at the Landau Institute in the Soviet Union and Michael Kosterlitz and David Thouless at the University of Birmingham showed that the unbinding of pairs of vortices was responsible for the transition to a disordered state in two dimensions. Thus, superconducting and superfluid transitions are qualitatively different in two and three dimensions. They are generally much smoother than the abrupt changes seen in three dimensions. Kosterlitz and Thouless were awarded a Nobel Prize in Physics in 2016.

The Woodstock of physics

I now turn to a class of quasi-two-dimensional superconducting materials. Their unexpected discovery in the late 1980s was one of the most exciting and surprising developments in condensed matter physics in the past 40 years. A 'holy grail' is to discover a material that is a superconductor at room temperature. This could revolutionize the transmission of electric power. Until 1986, the highest temperature at which superconductivity had been observed was 23 K (–250 °C), in Nb_3Ge, a combination of niobium and germanium. This requires cooling the material with liquid helium, which is expensive, and consequently limits commercial applications of superconductivity, to, for example, MRI machines in hospitals.

In 1986, Alex Bednorz and Karl Muller, working at an IBM laboratory in Switzerland, investigated whether a chemical compound composed of the chemical elements lanthanum, barium, copper, and oxygen would be a superconductor. They were motivated by theoretical arguments that strong interactions between the electrons and the vibrations of the atoms in the crystal could enhance the transition temperature, T_c, to the superconducting state. Bednorz and Muller observed superconductivity below a T_c of 36 K, a new record. The material consisted of layers of copper and oxygen atoms, and this stimulated many other research groups to investigate similar classes of chemical compounds. Within a year, materials were discovered with a T_c of 120 K (–150 °C). This was significant because superconductivity could be achieved by cooling with liquid nitrogen, which is much cheaper and more readily available than liquid helium. Unfortunately, over the past 30 years there has been little progress at increasing T_c to higher temperatures. Room temperature superconductivity remains a holy grail.

The discovery of Bednorz and Muller generated considerable excitement in the physics community, attracting many new researchers to the field of superconductivity. In March 1987 a special session was held during a regular meeting of the Division of Condensed Matter Physics of the American Physical Society in New York city. A large meeting room was overflowing with more than 1,000 physicists. I was one of thousands more outside watching on TV monitors. Each speaker in this special evening session was only allowed three minutes to present their work and the session went long past midnight. The event was described the following day on the front page of *The New York Times* as the 'Woodstock of Physics', after the famous rock music festival held in New York state that is considered emblematic of the 1960s.

The microscopic origin of the quasi-two-dimensionality of these materials is illustrated in Figure 22 (a). The layers of copper and oxygen atoms are in the centre of the repeat unit and are isolated from each other by a large number of other atoms. Consequently, it is much more difficult for electrons to move between the layers (the vertical direction in Figure 22 (a)) than in directions parallel to the layers. The structural and chemical complexity is reflected in the unit cell containing more than 50 atoms, including five different chemical elements. In contrast, in elemental lead there are only two atoms in the repeat unit (Figure 22 (b)).

An undergraduate student once asked me, 'what is the most interesting phase diagram in physics?' I drew the one shown in Figure 23. From the point of view of fundamental physics, the superconductivity of these materials is not the only property that is interesting or that presents a significant challenge to theoretical physicists. The materials exhibit two new states of matter, known as the pseudogap state and the strange metal state. These two conducting states have properties qualitatively different from common metals such as copper and gold. Theoretical models that successfully describe the latter fail for these new states.

Most crystals are either a metal, insulator, or semiconductor. These correspond to being a good, bad, or mediocre conductor of heat and electricity. In the pseudogap state electrons behave like those in a semiconductor, except when moving in certain directions within the crystal. In those directions they behave like electrons in a metal. In the strange metal state, the electrons act

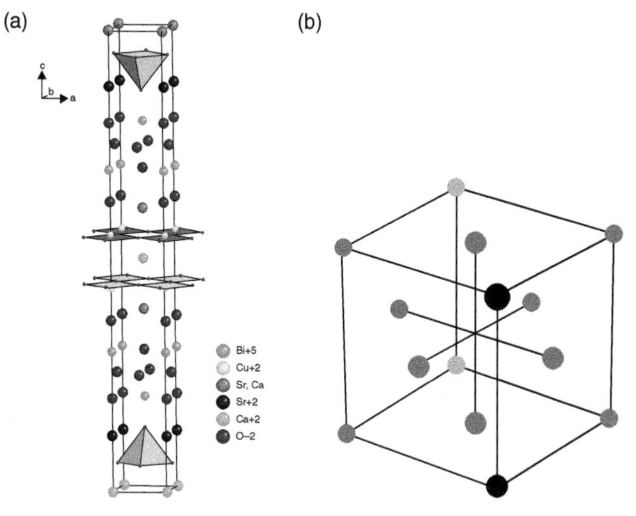

22. The chemical and structural complexity of copper oxide superconductors. (a) The repeat unit for the crystal structure of BiSrCaCuO. (b) The repeat unit for lead. It has only a few atoms in the repeat unit.

collectively in a manner that makes the concept of individual electrons meaningless. Since 1986 more than 10,000 scientific papers have been published concerning possible theories to describe the different states in this phase diagram. Although progress has been made, there is still no single theory that is accepted by the majority of physicists. In particular, there is no theory that has the simplicity and predictive power of the BCS theory of superconductivity (see Chapter 1) in simple metals such as lead and tin.

Both experimental and theoretical studies suggest that all this rich physics results from the layered structure of these copper oxide materials. The electrons are almost living in Flatland. There is no simple explanation for why this is so crucial, but generally as the spatial dimension of a system gets smaller, the components move

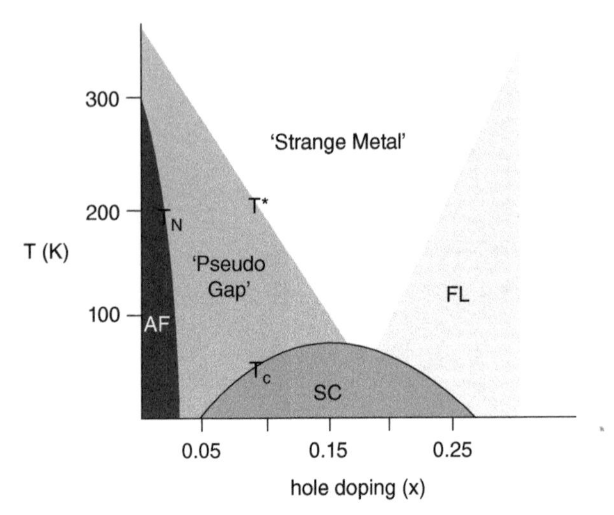

23. Phase diagram of copper oxide superconductors. Temperature (T) versus x, a parameter that describes the chemical composition of the material and is related to the density of charge carriers. The different states shown in the phase diagram are antiferromagnetic insulator (AF), superconductor (SC), regular metal (FL), strange metal, and pseudogap.

around more, fluctuations increase, ordered states are less stable, and new states of matter become possible.

The history of scientific discovery provides many examples of serendipity, where fortunate discoveries are unplanned. It is now understood that the reason that Bednorz and Muller chose to focus on this class of materials—a belief that there were strong interactions of electrons with atomic vibrations—was actually wrong. The high T_c does not arise from this interaction, but rather from a strong magnetic interaction between the electrons and from the two-dimensionality of the materials. Although the discovery was arguably somewhat serendipitous, its profound significance was highlighted by the fact that Bednorz and Muller were awarded the Nobel Prize only a year after their discovery.

The next chapter explores how, near a critical point in a phase diagram, changing the spatial dimension alters the properties of phase transitions. Flatland really is a different world. Furthermore, it is fruitful to think not only about systems with spatial dimensions of 1, 2, 3, and 4, but also fractional dimensions such as 3.999.

Chapter 6
The critical point

Every person is unique. No two people are identical. We differ in physical appearance, personality, fingerprints, heartbeat, gait, and DNA. Such differences are used to identify criminals and in video surveillance of citizens by nation states. Yet in other ways all humans are the same. We all have brains, hearts, and lungs. All our bodies use the same biochemistry to stay alive: whether to breathe oxygen, digest food, or fight infections. On some level we have common aspirations: to survive, to be loved, to be happy, and to find meaning and purpose. Yet these aspirations find many expressions. Humans have certain universal qualities and properties, yet at a finer level of detail there is a particularity of each of these properties. They are at one level the same but are not the same at another level.

All academic disciplines search for universals; they develop categories, concepts, and theories that overarch particularities. Biologists classify species of plants and animals and types of cells and viruses. All biological systems use the same molecules (DNA, RNA, and proteins) and chemical reactions. The same genetic code uses the information encoded in a piece of DNA to make proteins with specific functions. Anthropologists study the immense diversity of human cultures and societies. This diversity can be described in terms of universal concepts such as kinship, family, ritual, community, economics, law, and morality. Linguists study the common structures and grammars of the thousands of different human languages. Although the world we live in is diverse, disciplines have each discovered some universals.

Condensed matter physicists study diverse states of matter and the transitions between them. A surprising discovery is that there is much more universality than might be expected, particularly given the chemical and structural diversity of materials. In this chapter, I will discuss the nature of this universality, how it emerges, and the length scales associated with transitions between different states of matter. Landau's great insight was that many of the chemical and structural details of materials are irrelevant to understanding phase transitions. Furthermore, a precise classification of different types of phase transitions, into what are called universality classes, can be made. For example, superconducting, superfluid, and a subset of magnetic transitions are in the same class. The determinants of the universality classes are the symmetry of the state and the spatial dimensionality of the system. None of the other details matter.

Many phase diagrams (such as Figure 6) include a critical point, located at the end of a boundary between two different states of matter. A common example is the critical point that occurs at a specific temperature and pressure for a transition between a liquid and a gas. Understanding the physical properties of a material close to its critical point was a great challenge for theoretical physics, lasting 100 years, and was only solved in the 1970s. The powerful theoretical ideas and techniques that were developed provide a quantitative way to relate the properties of a system at one length scale to properties at a different length scale. These techniques also have application to a wide range of other problems and fields including elementary particle physics, chaos theory, fractals, polymers, and machine learning. New insights were gained into universality and emergent phenomena.

Universality near the liquid–gas critical point

For liquid–gas transitions the critical temperatures and the critical pressures for different pure substances, such as helium, argon, or water, vary from one substance to another, by a factor of more than 100. This large range arises because the interactions between the constituent atoms are different in the different materials. Nevertheless, close to the critical point, the curves mapping out the physical properties are identical in shape. Figure 24 shows an example with experimental data for the temperature dependence of the liquid and gas states for eight different pure substances.

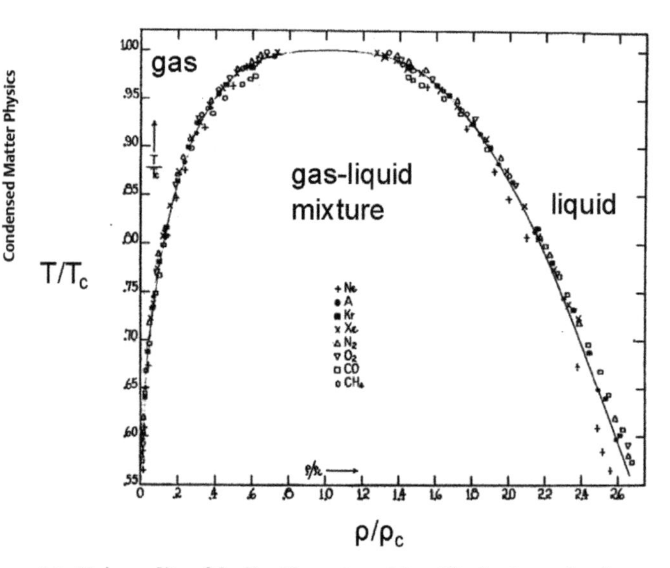

24. Universality of the liquid–gas transition. The horizontal and vertical axes are the density and the temperature, respectively, relative to their values at the critical point. Experimental data is shown for eight different materials (neon, argon, krypton, xenon, nitrogen, oxygen, carbon monoxide, and methane). The left (right) curve shows the relationship between the density and the temperature of the gaseous (liquid) state for temperatures below the critical temperature.

This graph was first published in 1945 by the physical chemist Edward A. Guggenheim and presented a challenge for theoretical physicists to explain.

Figure 24 shows that near the critical point the physical properties for different materials are in some sense the 'same'. The density and the temperature (or other relevant variables) for each material can be 'rescaled', that is, taken as a fraction of their value at the critical point. Then all the data, for a wide range of different materials, lies on a single curve. Understanding the *universality* associated with this scaling took several decades. The answer lies in thinking about rescaling the lengths in the system. Or to put it another way, we need to examine the differences in the appearance of the system as we zoom in and out.

Universality of critical exponents

The specific heat capacity of a material quantifies how much heat energy is required to increase the temperature of one gram of the material by one degree. For all critical points it is observed that

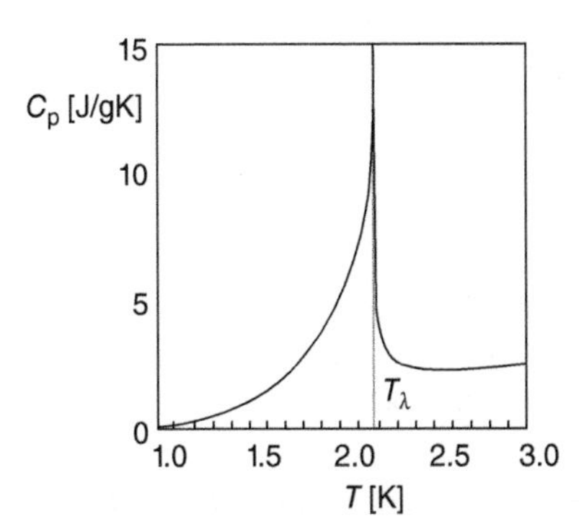

25. The temperature dependence of the specific heat capacity of liquid helium ^{4}He at temperatures several degrees above absolute zero. Near a temperature of 2.17 K the liquid undergoes a transition to a superfluid.

the specific heat becomes larger and larger as the material gets closer and closer to the critical temperature. The case of a superfluid transition is shown in Figure 25. The shape of the curve can be quantified in terms of a single number, denoted by the Greek letter α, referred to as the critical exponent for the specific heat. Surprisingly, the specific heat capacity of a large class of magnetic materials shows the same behaviour, close to the critical temperature for their magnetic transition, regardless of whether the material is ferromagnetic or antiferromagnetic, and the chemical composition and crystal structure of the material. The specific heat capacity has the same critical exponent as the superfluid transition. This is another example of universality.

The order parameter for a phase transition is zero at temperatures above the critical temperature and increases smoothly to non-zero values as the temperature is lowered below T_c. The mathematical form of this temperature dependence is specified by a critical exponent β. There are other critical exponents for other physical quantities, such as γ, which describes how the compressibility of a fluid gets larger and larger as the temperature gets closer to the critical temperature. The exponent δ describes the dependence of the density of a fluid on the pressure at the critical temperature, for a liquid–gas critical point.

Experiments on a wide range of materials and types of phase transitions show that there are universality classes of critical points. In a single class all the transitions have the same values for each of the critical exponents, regardless of the chemical composition and crystal structure of the material, or the magnitude of the critical temperature.

Here are some examples. For a whole family of ferromagnetic materials, such as iron, nickel, and cobalt, the critical temperature at which they lose magnetism varies by a factor of more than 100, from one material to another. The critical temperature for iron is

1043 K (770 °C) and for dysprosium is 88 K (–185 °C). Yet, the critical exponents for all these ferromagnets are the same, but different from those for the liquid–gas transition. The critical exponents for the liquid–gas transition, mixtures of two different liquids, metallic alloys composed of two different chemical elements (such as brass), and magnets described by the Ising model in three dimensions, are all the same.

Both experimental studies and theoretical work (discussed later) show that the universality classes are only determined by two numbers: n, the number of components of the order parameter, and d, the spatial dimensionality of the system. This is also where Flatland comes in. By making very thin films of a material, measurements can be made for a system with $d = 2$, and it is observed these systems have different critical exponents from those for $d = 3$.

While Landau recognized that many of the details of a phase transition are irrelevant to how physical properties, such as the specific heat, vary with temperature close to the critical temperature, he found that each of the critical exponents has the same value for all transitions. He found too much universality. The key missing ingredient in Landau's calculations turned out to be fluctuations; due to thermal jiggling a system does not have the same value of the order parameter at every spatial position and time. The collective effect of all this jiggling can be seen with the naked eye in a simple experiment.

Critical fluctuations

Charles Cagniard de la Tour discovered the critical point for liquid–gas transitions in 1822. A year later, he observed that a mixture of alcohol and water became milky at temperatures close to the critical temperature. The significance of this phenomenon, known as *critical opalescence* (Figure 26), was highlighted by Thomas Andrews in 1869 after he performed

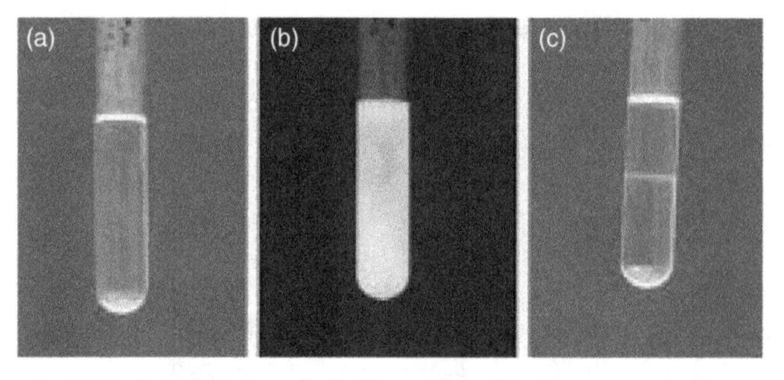

26. Critical opalescence of a fluid near the critical temperature for a liquid–gas transition. (a) The temperature is less than the critical temperature T_c and there are distinct liquid and gas states. The liquid has higher density than the gas and so is below the gas; the boundary between them can be seen. (b) The temperature is close to T_c and the entire fluid appears milky. (c) The temperature is larger than T_c and there is a single transparent fluid, known as a supercritical fluid; there is no gas–liquid boundary.

experiments on the liquid–gas transition in carbon dioxide. It was not until early in the 20th century that Marian Smoluchowski and Albert Einstein showed that the physical origin of critical opalescence is fluctuations in the density of the fluid. Close to the critical point these fluctuations become very large in magnitude and involve the formation of blobs of liquid and gas with sizes comparable to the wavelength of light. Consequently, light scatters off the fluid, causing it to appear opaque and milky.

For temperatures and pressures larger than at the critical point the system cannot be classified as a liquid or a gas, as can be seen in Figure 21 (a); it is called a supercritical fluid. The atmospheres of the planets Jupiter and Saturn are supercritical fluids. Supercritical fluids have commercial applications. Supercritical water is used in large power turbines because, unlike in regular steam, small liquid droplets cannot form, reducing the damage to power turbines. Supercritical carbon dioxide is central to the process used to decaffeinate coffee.

The size of the blobs of liquid and gas grows as the temperature approaches the critical temperature, reflecting a universal feature of all systems near critical points. In the case of magnetic systems, the blobs are magnetic domains, and this physics is captured in the Ising model. Compare the relative size of the black or white domains at the three different temperatures shown in Figure 21. At the critical temperature the size of the domains becomes as large as the system, reflecting the incipient order in the system. The emergence of large domains near the critical temperature was central to the theoretical breakthrough that provided a detailed understanding of critical phenomena.

Scaling

Suppose we consider an Ising model where all the lengths in the system are rescaled. In other words, if we zoom in and out, how do properties of the system change? In particular how does the strength of the interactions between the components and the size of the domains change? For example, suppose that a square lattice is divided up into 3×3 blocks, each containing 9 boxes. If the majority of the squares in a block is black (white) the block is replaced with a single black (white) box. Then repeat the process. The result of such a process is shown in Figure 27 when the temperature is equal to the critical temperature. The state of the system looks pretty much the same, regardless of the scale. In contrast, at temperatures above the critical temperature, the system scales to a completely random configuration of black and white squares. Below the critical temperature, the system eventually scales to a state of either completely black or white.

This approach was first investigated in 1966 by Leo Kadanoff. It was developed into a powerful mathematical form by Ken Wilson in the early 1970s and is known by an obscure name, the renormalization group. Wilson modified Landau's theory of the critical point to include the effects of fluctuations in the order parameter. He provided an explanation of universality and a

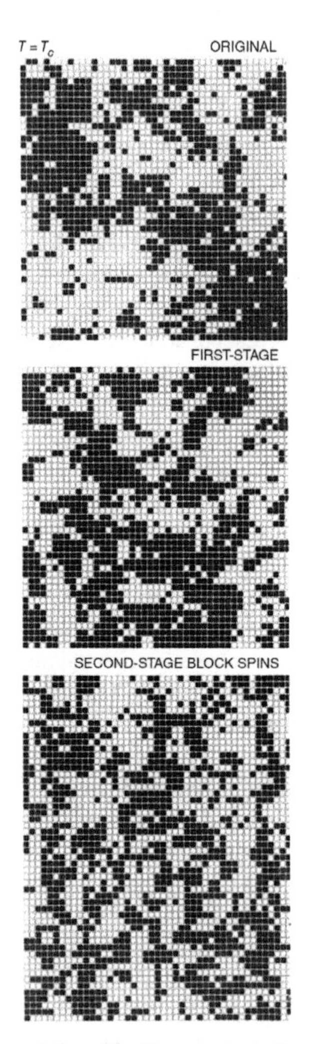

27. Scale invariance at the critical temperature. Rescaling of likely configurations of the Ising model on a square lattice. The state of the system looks essentially the same regardless of the rescaling.

method to accurately calculate the critical exponents of the different universality classes. The values obtained by Wilson's method were consistent with experiments and with computer simulations of Ising and related models.

Normally, we think of the dimension d of a spatial object, whether real or mathematical, as having integer values, $d = 1, 2, 3$. The size of an object is defined by length, area, and volume, respectively. In some mathematical calculations it is possible to consider the dimension d as having non-integer values, such as 2.5 or 3.874, and being a variable parameter. One can then investigate mathematically how the properties of the system change as the parameter d is slowly changed. Wilson and Michael Fisher embraced this idea. The theory of the critical point is simple to analyse when $d = 4$ (Landau's theory gives all the correct answers) but notoriously difficult when $d = 3$. A tractable way to find answers for $d = 3$ is do the calculations in 3.9999 dimensions and smaller dimensions and then use the results to extrapolate to $d = 3$.

Like Onsager, Wilson had an interesting publication and career history, very different from that expected of professors by universities today. Prior to working in condensed matter physics, he worked on elementary particle physics, completing a Ph.D. at Caltech with Murray Gell Mann. Wilson then worked at Harvard before taking a job as an assistant professor at Cornell University in 1963. He was widely viewed as brilliant. However, when Cornell had to decide whether he should receive tenure (i.e. have a permanent job until retirement), Wilson had only published a couple of articles in scientific journals, far fewer than would normally be expected. Nevertheless, Cornell took a risk and gave him tenure. This gamble paid off. In the early 1970s Wilson published a flood of papers that had a profound influence, not just on condensed matter physics, but also on the theory of elementary particles. Wilson was awarded the Nobel Prize in Physics in 1982.

Key to the successful development of Wilson's approach was that he took concepts and techniques that he had learnt from the theory of elementary particles and applied them in a new context. This was another example of the significant cross-fertilization between these two sub-fields of physics, which began in the 1950s and was crucial to the development of the BCS theory of superconductivity and of the standard model for elementary particles, including the prediction of the Higgs boson.

The renormalization group method developed by Wilson led to detailed calculations of the numerical value of the critical exponents associated with different universality classes. This in turn motivated further careful experimental measurements of the values of the critical exponents. The agreement between theory and experiment is impressive. Here is one example. Liquid helium becomes a superfluid below a temperature of 2.17 K. This is a critical point, and the specific heat becomes extremely large as the temperature approaches close to the critical temperature (Figure 25). A precise way to measure the critical exponents is to do the experiment in a zero-gravity environment, as this reduces variations in the density and pressure of the fluid between the top and the bottom of the container. In 1992 an experiment was performed on the space shuttle *Columbia*. Not only was the fluid cooled down to a temperature of a few degrees above absolute zero, but the temperature was controlled and measured to a precision of one-billionth of one degree. The specific heat capacity was measured and the corresponding critical exponent α was determined to have a value of –0.0127, that is, to five significant figures. This value was consistent with the value of –0.012 calculated from the renormalization group theory and computer simulations of the relevant lattice model, although there is still some debate about the last digit. The observed value of α disagrees with the value of exactly zero predicted by the simple theory of Landau, showing that he was quite close to the truth, but not exactly there.

Wilson's approach showed how universality emerges, that is, how systems with very different chemical and physical structures can have the same critical exponents. For example, consider the liquid–gas transition for water and the ordering transition associated with an alloy of copper and zinc. When these systems are viewed at the atomic scale, they are very different. However, as the critical point is approached the properties of the whole system are determined not by individual atoms or even a few atoms, but rather by collections of trillions of atoms (domains). The character of these domains and how they interact with one another are the same for systems in the same universality class; they are just determined by the symmetry of the order and the spatial dimensionality of the system. The atomic details do not matter.

This leads to the idea of 'effective theories', such as Landau's theory of phase transitions: at each length scale at which one looks at a system there is a theory that can describe the system at that scale. As one 'zooms out' or 'coarse grains' the system, theories often become much simpler, easier to understand and to investigate with mathematics or with computers. This important idea is related to emergence and is discussed in Chapter 9.

Systems in condensed matter involve a hierarchy of scales. At each scale there are corresponding phenomena, length scales, theories, and concepts. For example, ferromagnetism in a crystal of iron can be viewed at the following different scales: the individual electrons in each of the iron atoms, the magnetism associated with each atom, the magnetic interactions between the iron atoms located next to each other in the crystal, the magnetic domains in the ferromagnet, and the large size of these domains near the critical temperature.

Scale-free systems

At the critical point a system is scale free. It does not matter at what length scale the system is examined, it looks the same.

Beyond condensed matter physics, there are many other systems that are scale free. Fractals are examples (Figure 28).

Detailed quantitative analysis of data from a range of physical, biological, economic, and social systems shows that some of their properties show scale-free behaviour. An example is Zipf's law in linguistics: in any language there are only a few words that are used very frequently, while there are lots of words that are used far less frequently. Another example is Pareto's Principle concerning distribution of wealth (the 80/20 rule): 80 per cent of the wealth is owned by 20 per cent of the people.

Philosophers have debated the relationship between particulars and universals for centuries. In free societies there are public debates about values and morality; what is universal and what is particular to specific individuals, cultures, or situations? Academic disciplines have discovered certain universal patterns yet struggle to understand how universality does or does not emerge in the presence of particularity. This struggle is due to the complexity of the systems of interest, including the many different scales present. Condensed matter physics provides a concrete example where the relationship between universality and particularity is understood.

The next chapter explores how the quantum weirdness associated with the microscopic world of atoms, electrons, and light particles can also occur in the macroscopic world of states of matter, including superconductors and superfluids.

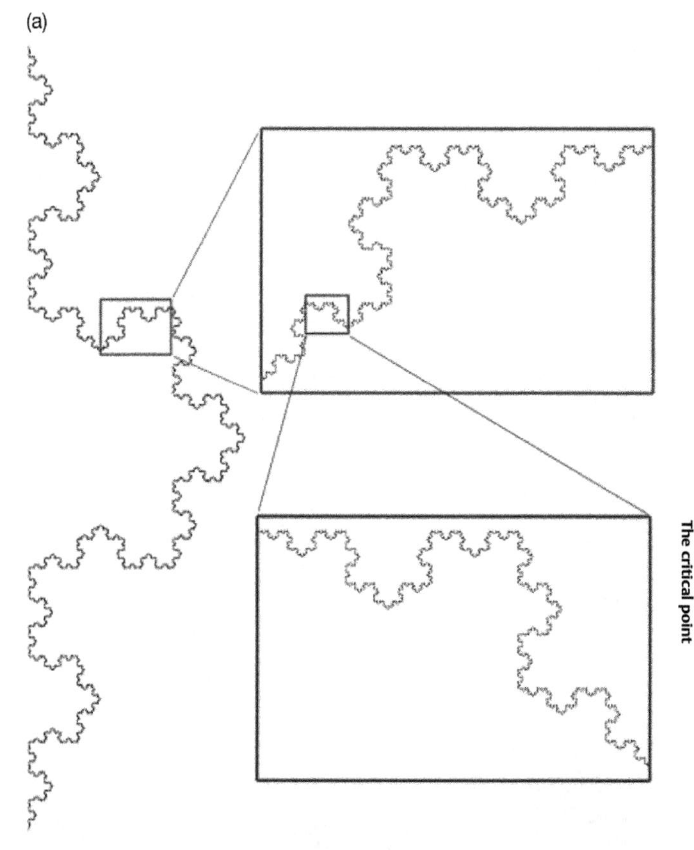

28. Fractals are scale-free patterns. When the image of part of the pattern (inside one of the boxes) is blown up the pattern still looks the same. Proceeding from left to right and then top to bottom one sees that the pattern is the same, for both the (a) Koch snowflake and (b) Julia set. For the latter, at each step the scale of the picture is increased by a factor of four. The final picture (bottom right) covers an area of about one millionth of the original picture (top left).

(b)

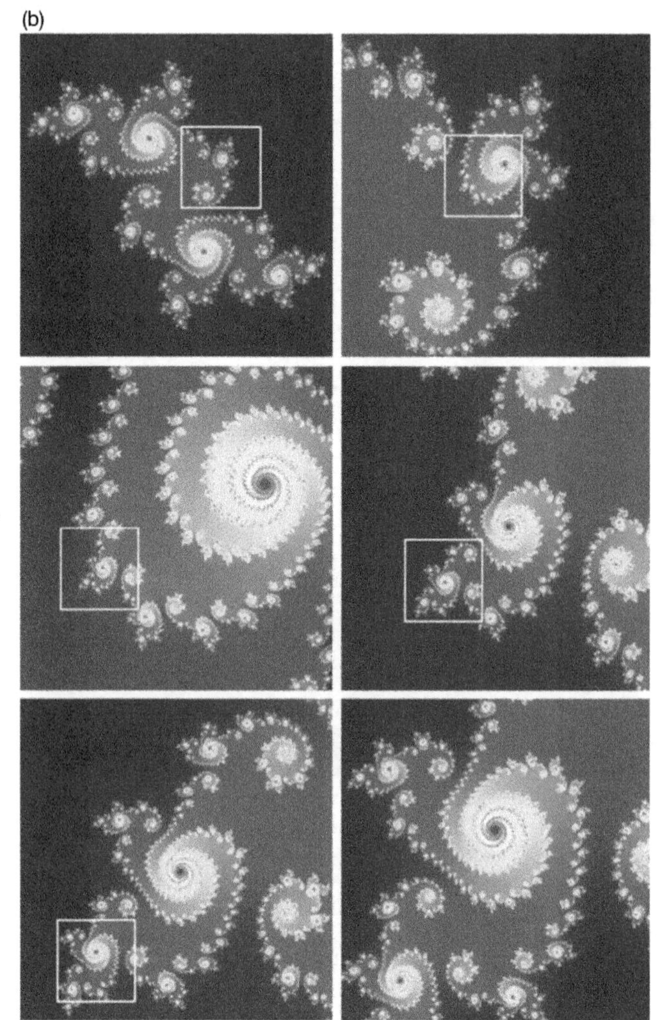

28. Continued

Chapter 7
Quantum matter

At the end of the 19th century, physicists were settled on their view of the world. Newton's laws of motion and of gravity were a great success, whether describing the motion of the planets or of projectiles. It was well established that particles (e.g. a football) and waves (e.g. light or ripples on the surface of a pond) were distinct entities exhibiting distinct phenomena. A particle could not be in two places at the same time. The position and speed of a particle could both be precisely measured at the same time. The theory of electromagnetism developed by James Clerk Maxwell in the 1860s gave a unified and complete description of electricity, magnetism, and light waves. However, the birth of quantum theory in the first three decades of the 20th century turned upside down physicists' view of the world. The quantum world of atoms, electrons, protons, and photons (light particles) was weird and very different from everyday experience.

Erwin Schrödinger was one of the founders of quantum theory. The mathematical equation that bears his name, published in 1926, is the foundation of the quantitative description of atoms, molecules, and crystals. Due to the rise of the Nazi Party, Schrödinger left Germany in 1933 and took a temporary job at Oxford University. For a while he lived in the house next door to J. R. R. Tolkien, author of the *Lord of the Rings*. Perhaps Tolkien provided some inspiration for a highly imaginative scientific paper that Schrödinger wrote in 1935. Schrödinger did acknowledge the paper was written in response to criticisms of quantum theory by Albert Einstein. Schrödinger's paper described a 'thought experiment' involving a cat, to illustrate the 'quite ridiculous cases' that are possible with quantum theory.

Schrödinger imagined placing a cat in a box together with a small container of radioactive atoms and a glass vial of poison. The radioactive atoms have a long half-life and there are so few of them in the container that on average only about one atom decays per hour. When an atom undergoes radioactive decay, it would emit a gamma ray, detected by a Geiger counter that would trigger a hammer to smash the glass vial, releasing the poison, and killing the cat. According to quantum theory individual atoms decay at random times and it is possible to have a situation where an atom is simultaneously in two states, decayed and not decayed. Thus, it is possible to have the cat simultaneously dead and alive! Furthermore, in quantum theory any measurement forces a system to have a definite state. Hence, if after an hour someone opens the lid of the box and looks inside then the cat must adopt one of the two possible states, dead or alive.

Over the past 100 years quantum theory has undergone numerous experimental tests on a diverse range of systems, from electrons to molecules. These tests have been of such high precision that quantum theory is arguably the most successful theory in science. Yet, the paradox of Schrödinger's cat has never been resolved, at least not to the satisfaction of most physicists and philosophers. A key issue is the location of the boundary between the everyday *macroscopic* world of definite outcomes (cats must be either dead or alive) and the *microscopic* quantum world of atoms where outcomes are indeterminate. For example, as the number of atoms in a molecule is increased, when does the molecule become so large that it starts to act like an everyday object?

This chapter explores how some states of matter, such as superconductors and superfluids, can manifest quantum weirdness at the macroscopic scale. This is of fundamental scientific and philosophical interest. It is also the basis for some new technologies, highly sensitive scientific instruments, and precise electrical measurements.

Quantum steps

When you drive a car, you can vary the speed of the car continuously by smoothly varying the pressure you apply to the accelerator pedal. As a result, the speed of the car can have any value, between zero and its top speed. The speed of the car and its energy of motion does not jump between discrete values. The world of atoms and electrons is not like that. It is much lumpier. The energy of an atomic system cannot have any possible value, only certain discrete values. Energy comes in discrete lumps or quanta. It is quantized. The difference between these allowed values of the energy is extremely small compared to the energy of everyday objects. For the motion of a football, we could never see the discreteness in practice. It is like how when you look at a high-resolution photo you cannot see the individual pixels, or a staircase where the height of the steps is so small that it looks like a smooth slope. In contrast, the energies of an electron in a single atom are so small that the discrete differences really do matter. In quantum theory, not only energy but also other physical quantities such as momentum, angular momentum, wavelength, and magnetic flux can be quantized.

Quantum interference

If you throw two pebbles into a still pond, you may observe an interference pattern between the two sets of waves emanating from the points on the pond surface at which the two pebbles hit the water. At some points on the surface of the pond the amplitude of the wave is larger and in some parts it is smaller.

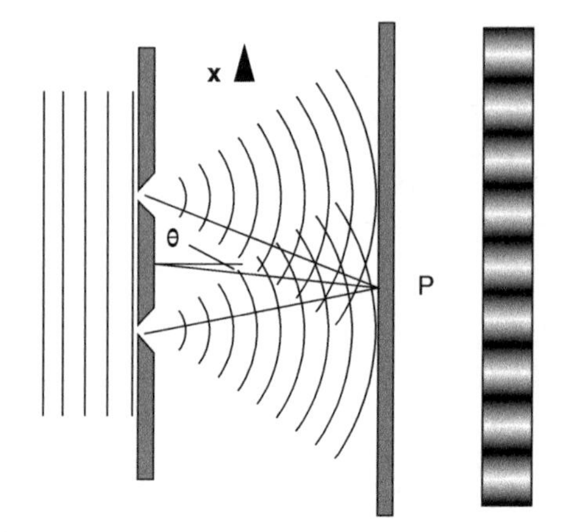

29. Double-slit experiment showing wave interference. A beam of light is shone from the left onto a screen which contains two narrow slits. The thin narrow lines represent crests of the wave. Each slit acts as a source of separate new waves. These two waves interfere with one another so that on the screen P an alternating pattern of light and dark is seen.

Similar wave interference is involved in the X-ray diffraction patterns produced by crystals and used to determine their atomic structure.

Isaac Newton believed that light was composed of particles. Others believed light was a wave. In 1801 Thomas Young performed an important experiment to establish that light was a wave (Figure 29).

In 1927 Clinton Davisson and Lester Germer performed a version of Young's double-slit experiment in which they replaced the beam of light with a beam of electrons. They found that the intensity of the electron beam observed on the screen showed an interference pattern just like that seen in Young's experiment with light. The electrons were apparently behaving as waves. This raises questions connected to the paradox of Schrödinger's cat and that physicists still debate. Which of the two slits does a single electron pass through? Or does it pass through both slits simultaneously. Is the electron in two places at the same time? Is the electron a wave or a particle? It was also discovered that light sometimes acts like a particle, called a photon. To describe this strange situation, Niels Bohr coined the term 'wave–particle duality'. Electrons and photons sometimes act like particles and sometimes like waves; it depends on the context.

Regarding the location of the boundary between the everyday world and the microscopic world of quantum weirdness, the question is how large must a particle be in order to see quantum interference? In the past two decades impressive double-slit experiments have been performed using beams of large molecules that have a mass that is about 10,000 times larger than the mass of a single hydrogen atom. If there is a boundary physicists have not found it yet.

Fundamental constants

Since the beginning of the 20th century, physicists have made many stunning discoveries. The speed of light is constant. It is always the same regardless of where it is measured, by whom, or when, or how fast the observer is moving. The most accurate measured value is 299,792,458 metres per second, and the symbol c is used to denote the speed of light. Every electron in the universe is identical. They all have the same electrical charge and same mass. The charge of an electron is denoted by the symbol e and has the value $e = 1.60217662 \times 10^{-19}$ coulombs. Planck's constant, denoted h, describes the proportionality between the energy of a photon and the frequency of oscillation, ν, of the associated wave, $E = h\nu$. The separation of allowed energy values is determined by h and it has the value of 6.6314×10^{-34} joule seconds. The quantities c, e, and h are all referred to as fundamental constants.

George Gamow was a playful theoretical physicist who wrote several popular books about science. Many are still in print. In *Mr Tompkins in Wonderland*, first published in 1939, Gamow took a non-scientist, Mr Tompkins, on a tour of an imaginary world in which the speed of light was one billion times smaller than in the real world and Planck's constant was so large that quantum effects could be seen in everyday life. In a dream, Mr Tompkins observes quantum interference on a macroscopic scale. While on an animal safari, a single gazelle passes through a bamboo grove. The grove and gazelle act like an electron in a multiple-slit experiment producing interfering waves. To Mr Tompkins there appear to be many gazelles. This is because the single gazelle is at many different locations simultaneously. This makes it difficult for a hunter to shoot the gazelle. Although Mr Tompkins saw macroscopic quantum effects only in his dreams, in Gamow's lifetime it became possible to observe them in the laboratory.

Macroscopic quantum effects in superconductors

Magnets and electrical currents produce magnetic fields, regions of space where other magnets and electrical wires experience a mechanical force. For a circle of wire in the presence of a magnetic field the magnetic flux is defined as the strength of the magnetic field passing through the circle multiplied by the area of the circle. A law of electromagnetism states that if the field varies with time, then a voltage is produced in the wire with a magnitude that is proportional to the rate at which the magnetic flux through the circle changes. This is the physics behind all electrical motors and electrical generators. In the everyday world magnetic flux can have any value and can be varied continuously by changing the strength of magnetic field. In the quantum world that is not the case. Magnetic flux is quantized.

In 1961, two experimental groups independently reported the first observation of a macroscopic quantum effect, the quantization of

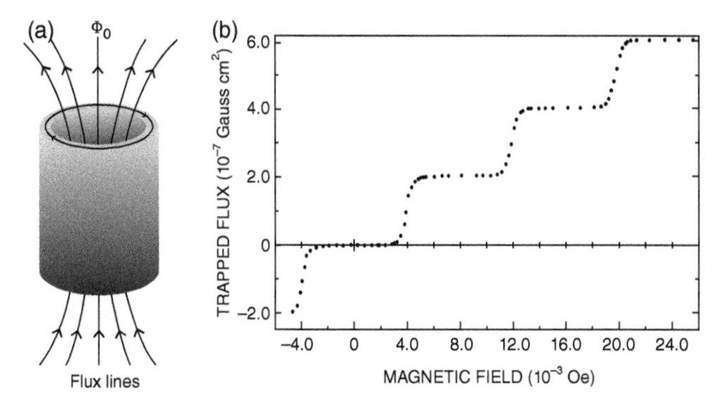

30. Quantization of magnetic flux in a superconducting cylinder. (a) A tall thin cylinder of tin was placed in a magnetic field. (b) The graph shows the value of the magnetic flux passing through the cylinder as the magnetic field was varied. Note the step-like structure, showing quantization of the flux.

the magnetic flux passing through a superconducting cylinder (Figure 30). One team was Bascom Deaver and William Fairbank and the other Robert Doll and Martin Nabauer. A tall thin cylinder made of tin was placed in a magnetic field and cooled down to a low enough temperature that it entered the superconducting state. The magnetic flux passing through the cylinder was then measured as the magnetic field was varied. The resulting graph has four noteworthy features. First, there are clear steps, showing that the magnetic flux has discrete values. In contrast, in the normal metallic state the graph was a straight line. Secondly, the magnitude of the steps was the same, to within about 1 per cent, suggesting quantization of a single unit of magnetic flux. Thirdly, the value of this quantum of magnetic flux was equal to the value of $h/2e$. Thus, it was completely determined by the two fundamental constants, h and e, Planck's constant and the charge on an electron, respectively. And fourthly, graphs with the same three features noted above were later observed in other superconducting materials and cylinders. This showed that flux quantization is independent of details such as the chemical composition and dimensions of the cylinder. This flux quantization is a macroscopic quantum effect. It is macroscopic because the system is macroscopic, and the magnetic flux is a macroscopic property. It is a quantum effect because the magnitude of the quantization is determined by Planck's constant.

The quantum of magnetic flux is denoted Φ_0 ($= h/2e$) and has the value $2.067833848\ldots\times10^{-15}$ tesla (metre)^2. This number also determines the scale of quantum interference effects between two superconductors, as we will see shortly. The flux quantum is also relevant to vortices that form when some superconductors are placed in a magnetic field (Figure 20). A persistent electrical current flows around the vortex and the magnetic field penetrates the core of the vortex. It can also be shown, both theoretically and experimentally, that the magnetic flux associated with each vortex is exactly equal to one quantum of flux. Something similar happens in superfluids.

Macroscopic quantum effects in superfluids

When a cylinder containing a fluid is rotated about an axis passing down the centre of the cylinder the fluid will also rotate. The faster the cylinder is rotated the faster the fluid rotates. A physical quantity known as the circulation is proportional to the speed of rotation and the diameter of the cylinder. With a variable speed motor, the rotation speed can be continuously varied and in normal fluids the circulation has continuous values. But not in a superfluid, as shown in a beautiful experiment done by W. F. Vinen in 1961 using liquid ^{4}He. He observed that when the liquid was cooled below the superfluid transition temperature the circulation could only take on discrete values. Furthermore, these discrete values are multiples of h/M where h is Planck's constant and M is the mass of one atom of helium. This value was predicted by Lars Onsager in 1949 who identified h/M with the circulation of a single vortex in the superfluid. This is another macroscopic quantum effect.

The quantization of magnetic flux in superconductors and of circulation in superfluids showed that both superconductors and superfluids can be classified as quantum states of matter. These quantum phenomena are quite similar, even though superconductivity occurs in solids and superfluidity in liquids. This indicates a deep underlying unity, demonstrated through the study of condensed matter physics.

Josephson junctions

The observation of macroscopic quantum effects made another advance at Cambridge University in the early 1960s. When the theory of superconductivity was developed in 1957 by Bardeen, Cooper, and Schrieffer (BCS) they did not fully appreciate its connection with the Ginzburg–Landau theory of superconductivity, including the physical significance of the order parameter and spontaneous symmetry breaking. In a superconductor the value of the order parameter at any point in space is two numbers, one of which can be viewed as the angle that defines the state in terms of a point on the circle at the bottom of the Mexican hat potential (Figure 18). While spending a sabbatical year at Cambridge, Phil Anderson gave a series of lectures on the theory of solids and discussed the importance of the concept of broken symmetry. One of the Ph.D. students who attended Anderson's lectures was Brian Josephson. He began to think about possible experiments to measure the angle associated with the order parameter for superconductivity.

In 1962, Josephson proposed investigation of the electrical circuit shown in Figure 31. The central feature is the 'junction' where a very thin layer of an insulating material is wedged between two layers of superconducting material. At temperatures higher than the superconducting critical temperature these two layers are normal metals and when the circuit is not connected to a battery there is no electrical current in the circuit. Josephson argued that

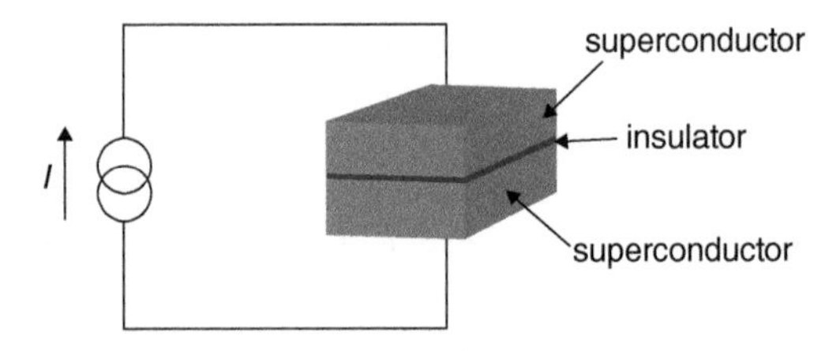

31. A Josephson junction in an electrical circuit. A current, denoted *I*, flows around the circuit passing through a very thin layer of an insulator that is wedged between two superconductors. The magnitude of the electrical current depends on differences between the values of the order parameters of the two superconductors.

when the two metals were in the superconducting state there could be a current, even without a battery, and the size of the current would depend on the difference between the angles associated with the order parameters of the two superconducting states.

Josephson's proposal was rejected and actively opposed by John Bardeen, a widely respected figure. Bardeen had been awarded a Nobel Prize in 1956 for the discovery of the transistor and led the team that created the BCS theory of superconductivity in 1957. Nevertheless, Josephson held firm and tried to construct the electrical circuit that he proposed. Although he did not succeed, on Anderson's return to Bell Labs in the USA, he and John Rowell were able to create the circuit and observe the predicted effects. Josephson shared the Nobel Prize in Physics in 1973.

Josephson considered a beam of microwaves with a single frequency incident on a junction. He predicted that there would be jumps in the current passing through the junction as the voltage across the junction was varied. These jumps would occur when the voltage was equal to some integer multiple of the microwave frequency times $h/2e$. This is known as the AC

Josephson effect, where AC stands for the alternating electrical current associated with generating the microwave.

It is possible to measure voltages and frequencies with extremely high precision and so the AC Josephson effect enables precise measurement of the ratio h/e. Prior to Josephson, measurement of such fundamental constants was only possible by measurements of the properties of single atoms and electrons. In contrast, Josephson provided a way to measure these microscopic parameters with a macroscopic system.

The AC Josephson effect is another example of a macroscopic quantum effect, with the characteristic four features, noted earlier for the quantization of magnetic flux: it is a macroscopic system in which a physical quantity has discrete, or stepped, values; the magnitude of the steps is an integer multiple of a parameter; that parameter is determined by fundamental constants, including Planck's constant; and the quantization is observed in different materials and devices.

The AC Josephson effect has an important application in metrology, the study of measurement and the associated units and standards. Prior to 1990 the international standard used to define one volt was based on a particular type of electrical battery, known as a Weston cell. In 1990 an international agreement was made to use the AC Josephson effect as the standard instead. This allowed voltages to be defined with a precision of better than one part per billion. This change was motivated not only by improved precision, but also portability, reproducibility, and flexibility. The old voltage standard involved a specific material and device and required making duplicate copies of the standard Weston cell. In contrast, the Josephson voltage standard is independent of the specific materials used and the details of the device. This independence reflects the fact that the Josephson effects have the universality characteristic of emergent phenomena.

Schrödinger's cat is a squid

Josephson junctions can exhibit quantum interference effects like the double-slit experiment and can be used to make extremely precise measurements of magnetic fields. Both involve a SQUID, a Superconducting Quantum Interference Device. This consists of a circuit in which an electrical current can pass through either of two Josephson junctions (Figure 32). Quantum interference occurs between the two currents. If a magnetic field passes through the loop the interference changes by an amount that depends on the value of the magnetic field relative to the quantum of magnetic flux. If the current is measured as a function of the strength of the magnetic field, then the current oscillates between a maximum value and zero, similar to the interference pattern for a double slit shown in Figure 29.

A SQUID can be used to make highly sensitive measurements of magnetic flux in laboratories and in diverse technological contexts. Indeed, SQUIDs are the most sensitive detectors of magnetic flux available. They can sense a change as small as one billionth of the Earth's magnetic field. SQUIDs are used in

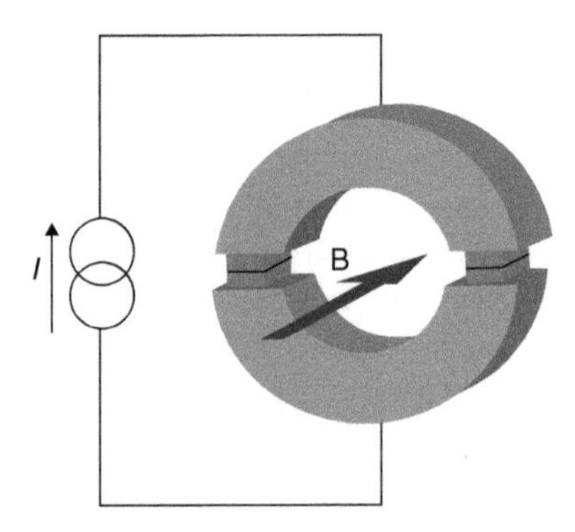

32. A superconducting quantum interference device (SQUID). Two Josephson junctions are part of a circuit loop that is placed inside another circuit. A magnetic field, denoted by B, passes through the loop.

geological prospecting, and to locate defects in silicon microchips and microscopic cracks in aircraft wings. In neuroscience, SQUIDs are used to detect the small magnetic fields produced by the electrical currents associated with brain activity and can help identify neurological disorders.

Since 2000, SQUIDs have been used to create quantum states just like Schrödinger's cat, where a macroscopic system is simultaneously in two different states. These states exhibit quantum weirdness such as tunnelling, interference, and entanglement, the strangest property of quantum theory. Entanglement is central to quantum computation, whereby the power of a computer is enhanced by the 'parallel processing' made possible by quantum weirdness. In principle this could make possible the performance of tasks such as cracking codes not possible with a traditional computer. However, construction of such a computer is a formidable challenge. So far, the most promising and powerful devices involve circuits containing tens of SQUIDs.

These SQUIDs are sometimes called 'artificial atoms' because they exhibit all the quantum weirdness associated with single atoms. In the early 1980s Tony Leggett proposed using SQUIDs to construct such systems and test whether macroscopic systems really were described by quantum theory. There was scepticism towards these proposals, particularly from experts working on the foundations of quantum theory. Leggett recently observed, 'I find it rather amusing that nowadays the younger generation of experimentalists... blithely writes papers with words like "artificial atom" in their titles, apparently unconscious of how controversial that claim once was.'

In this chapter we have seen that superconductors and superfluids are states of matter that are fundamentally different because they manifest quantum properties at the macroscopic scale. The next chapter considers another class of quantum states of matter that occur in Flatland. For these, the quantum steps can be related to the shapes of doughnuts and pretzels.

Chapter 8
Topology matters

Topology, as we noted earlier, is the field of mathematics describing the properties of geometric objects that do not change when they are smoothly deformed. These properties only change in steps by cutting or gluing. Concepts in topology can be illustrated with everyday objects such as balls, doughnuts, coffee cups, and pretzels. For example, a doughnut can be gradually and smoothly deformed into the shape of a coffee mug (Figure 33). No ripping or cutting is required. In contrast, it is impossible to turn a ball into a doughnut without cutting a hole. The number of holes in an object is referred as a topological invariant. For a ball, doughnut, and the simplest pretzel these numbers are zero, one, and two, respectively. Topology is about qualitative differences not quantitative details such as distances, angles, and sizes.

In Chapter 4 it was noted that in ordered states of matter, some properties are determined by topological defects, such as vortices in superconductors. These are topological objects in the following sense. In a superconductor, there is an electrical current circulating around a vortex and a magnetic field that passes through the centre of the vortex. The total magnetic flux through a vortex is equal to exactly one quantum of magnetic flux. If the spatial distribution of the electrical current around the vortex is smoothly changed the total magnetic flux remains the same. Furthermore, it is not possible to smoothly deform the system in any way to make the

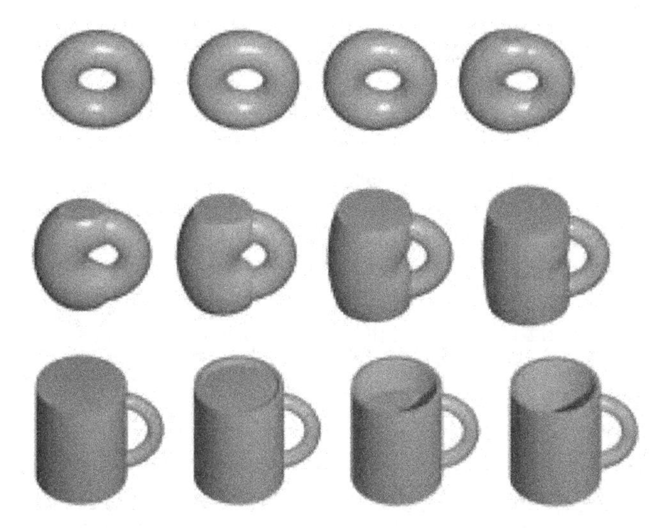

33. A doughnut can be smoothly deformed into a coffee cup. From the perspective of the mathematical field of topology all the objects above are identical.

vortex disappear. The magnetic flux associated with the vortex is a topological invariant.

Condensed matter physics is about qualitative difference: states of matter are qualitatively different from one another. Until the 1980s these differences were only associated with different types of symmetry, which in turn reflect the underlying ordering in the state. This chapter describes unanticipated discoveries of new states of matter that could not be described in terms of this traditional symmetry picture. But they can be described in terms of topology. These states exhibit macroscopic quantum effects, reminiscent of superconductors and superfluids. Understanding these states of matter involves venturing back into Flatland and also into some abstract mathematical spaces. Remarkably, these abstractions can be related to practical questions about international standards for electronic circuits.

Classical Hall effect

The Hall effect in metals was discovered in 1879 by Edwin Hall. A voltage is applied across the ends of a thin metal strip so that an electrical current flows along the strip. When a magnetic field is present with a direction perpendicular to the plane of the strip a voltage (electrical potential energy difference) develops between the sides of the strip (Figure 34). This is known as the Hall voltage, and the Hall resistance is defined as the ratio of the Hall voltage to the magnitude of the electrical current that travels along the strip. The magnitude of the Hall resistance is proportional to the strength of the magnetic field and is inversely proportional to the density of charge carriers in the metal. Quantum theory is not needed to understand these properties of the Hall effect. It can be understood in terms of how a magnetic field exerts a force on a moving charged particle, so it does not travel in a straight line.

The Hall effect is used to investigate properties of metals and semiconductors because it allows determination of the density of charge carriers responsible for electrical conduction. But some

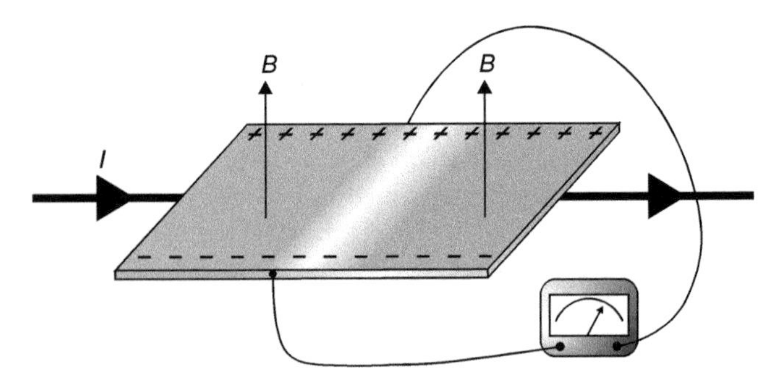

34. The Hall effect. A thin metallic strip is placed in a magnetic field that is perpendicular to the surface of the strip, and denoted B. An electrical current flowing along the strip leads to a voltage difference between the sides of the strip.

measurements presented a challenge to classical theories. Whether the Hall voltage was positive or negative varied from one material to another. This was inconsistent with the accepted picture that in metal electrical currents were due to the flow of electrons, known to have a negative electrical charge. This suggested that in some metals (e.g. magnesium and aluminium) the current was carried by particles with positive charge. In the 1930s this puzzle was solved by using quantum theory to understand the electronic properties of crystals. It was shown how positive charge carriers can naturally arise in certain crystals. These charge carriers are known as *holes* because they represent the absence of an electron in a 'sea' of many electrons. Holes are central to modern semiconductor technology. Diodes, transistors, and photovoltaic cells consist of interfaces between two regions. On one side of the interface electrical conduction occurs via electrons and on the other side via holes.

Landau levels

In a magnetic field, charged particles, such as electrons, move in circular orbits in a plane that is perpendicular to the direction of the magnetic field. However, this classical picture breaks down in quantum theory. Particles are like waves and so if they undergo circular motion, they will experience wave interference depending on the size of the circular orbit. Similar physics is responsible for the quantization of the energy of electrons in atoms.

In 1930, just a few years after Schrödinger wrote down his famous mathematical equation for quantum theory, a 22-year-old Lev Landau solved the equation for the wave motion of a charged particle in a magnetic field. He showed that the energy of the particle could only have certain discrete values, now known as Landau levels. There is an integer n (1, 2, 3, ...) associated with each of these quantum states, with the energy increasing with n. Landau noted that this discreteness of the energy could lead to the magnetic properties of a metal oscillating as the magnetic field was varied. However, Landau suggested these quantum effects were unlikely to be observed. In order for the electrons to be in the low-lying energy levels and complete an orbit without being scattered by impurities in the metal three criteria must be met: low temperatures, high magnetic fields, and pure materials. But, in fact at around the same time experimentalists did observe the oscillations that Landau studied theoretically. One of these scientists was Lev Shubnikov, who was later falsely accused and executed in one of Stalin's purges. This led to Landau writing the anti-Stalin pamphlet that resulted in his own arrest and imprisonment.

Integer quantum Hall effect

More surprises occurred in the 1980s when it became possible to study Landau levels in Flatland. This happens when the electrons are completely constrained to move in only two dimensions. The surface within which the electrons move needs to be extremely flat and free from defects and impurities. Advances in semiconductor technology in the 1970s led to two realizations of this Flatland. Both were developed for technological reasons: the desire to have transistors in which the electrons and holes can move extremely fast. One class of device is silicon MOSFETs (Metal Oxide Semiconductor Field Effect Transistors). The second class is heterostructures, where layers of ultrapure semiconductors such as gallium arsenide are grown on top of each other, one layer of atoms at a time. In both classes of device, a fixed density of electrons (or holes) can be injected at the surface. These charge carriers can move freely in Flatland, acting like a fluid. Things get interesting when the number of charge carriers is small enough and the magnetic field is large enough that the number of charge carriers is comparable to the number of quanta of magnetic flux that pass through the system. Then, the quantum state of most of the charge carriers is one of the lowest Landau energy levels.

To achieve this regime for the cleanest possible systems requires magnetic fields more than 100,000 times stronger than that of the Earth. Furthermore, the magnetic field must be spatially uniform in the region where the semiconductor system is located, stable over the time of the measurements, and the interior of the electromagnet producing the field must be large enough to contain a refrigerator that can cool the charge carriers in the system down to a few degrees above absolute zero. By 1980, all these conditions became possible. Klaus von Klitzing was able to perform measurements of the Hall resistance versus magnetic field in a special high magnetic field laboratory in Grenoble, France. The results were surprising and are shown schematically in Figure 35. There are four noteworthy features.

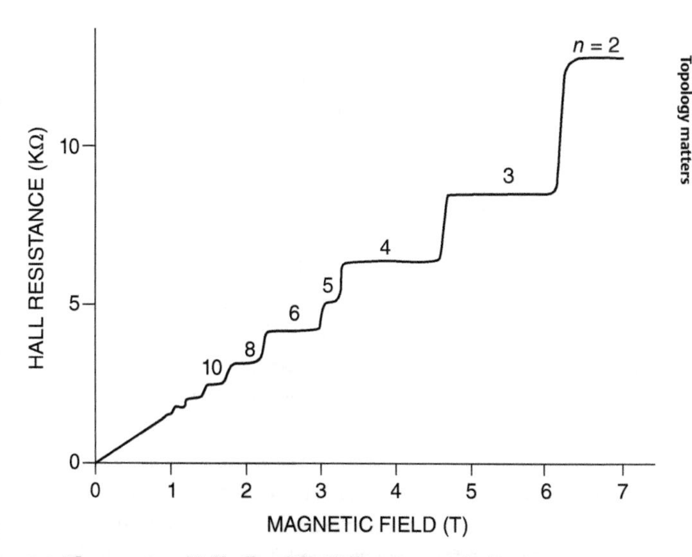

35. The quantum Hall effect. The Hall resistance is shown as a function of the strength of the magnetic field and has a step-like structure. The integer *n* is related to the quantized energy that the charge carriers have.

First, there are distinct steps in the curve. At small magnetic fields the Hall resistance versus field is a straight line, as expected for the classical Hall effect. However, at larger fields there are plateaus in the curve.

Second, each of the plateaus is extremely flat. Von Klitzing found that the magnitude of the Hall voltage on each plateau did not vary to one part in 10 million. As he varied the magnetic field, he noticed that the first seven digits on the voltmeter he was using did not change. He wondered if the voltmeter was broken and had become jammed. But it was working.

Third, the magnitude of the Hall resistance for all the plateaus has a simple relationship to fundamental physical constants. The quantum of resistance is defined as equal to $h/2e^2$. When you calculate this quantity, the answer (25,812.827 ohms) is in the units of electrical resistance. The value of the Hall resistance is precisely equal to this value divided by an integer ($n = 1, 2, 3\ldots$) which is related to the highest quantized energy (Landau level) that an electron can have at that magnetic field. That is why it is known as the integer quantum Hall effect.

Fourth, the observed value of the Hall resistance for each of the plateaus is independent of many details, including the temperature, the amount of disorder in the material, the chemical composition of system (silicon versus gallium arsenide), or whether the charge carriers are electrons or holes.

These four features are similar to those for the steps associated with the macroscopic quantum effects (magnetic flux in superconducting cylinders, circulation in a superfluid, Josephson effects) discussed in the previous chapter. Again, it is astonishing that a macroscopic measurement—of electrical resistance—of a macroscopic system can determine fundamental constants that are normally associated with properties of atomic systems. Just as the Josephson effect led to a new standard measure for voltage, the quantum Hall effect led to a new standard measure for electrical resistance.

Anyone familiar with building electronic circuits will have used resistors of varying values in ohms (Ω), such as 10 Ω or 25 kΩ. When these resistors are made, they are calibrated against some standard. For making integrated circuits with billions of transistors this standard needs to be extremely accurate. In 1990 the international standard for the ohm was changed to be that defined by the quantum Hall effect. Previously, the ohm was defined by the electrical resistance of a column of liquid mercury with constant cross-sectional area, 106.3 cm long, a mass of 14.4521 grams, and a temperature 0 °C. Like the Josephson voltage standard, the quantum Hall resistance standard has the advantage of precision, portability, reliability, reproducibility, and independence of platform.

From the practical to the abstract

Why is the Hall resistance so precisely quantized and independent of so many details? Topology provides the answer: the Hall resistance is a topological invariant. This insight came from a group of theorists led by David Thouless. The topology involved is not that of physical objects in the real three-dimensional space of the laboratory, but rather an abstract mathematical space that describes the quantum states that the electrons are in. Thouless, Mahito Kohmoto, Peter Nightingale, and Marcel den Nijs considered an abstract model in which the electrons move in the presence of a uniform magnetic field and a periodic electrical potential, such as occurs due to the arrangement of atomic ions in a two-dimensional crystal.

This model leads to beautiful patterns such as shown in Figure 36, a graph of the dependence of the allowed energy values on the magnetic flux. The resulting pattern has a fractal (scale-free) structure.

36. Hofstadter's butterfly. The allowed energies are shown as a function of magnetic field (flux) for charged particles moving in the presence of a two-dimensional periodic electric potential and a uniform magnetic field. The horizontal scale is related to how many quanta of magnetic flux pass through each repeat unit of the periodic potential. The vertical scale is related to how easy it is for the charge particles to move between neighbouring repeat units.

These patterns arise because there is competition between the periodicity of the crystal and the periodicity associated with the effect of the magnetic field on the wave nature of the electrons. If the two periodicities are commensurate then a continuous range of values for the energy of the electron is allowed (note the vertical lines at certain values of the magnetic flux). The model associated with these patterns was the subject of the Ph.D. thesis of Douglas Hofstadter, who became famous for his Pulitzer Prize winning book *Gödel, Escher, and Bach*, in which he reflected on patterns in mathematics, art, and music, and the insights these patterns give to human cognition and the emergence of consciousness.

The mathematical formula that Thouless's group derived for the Hall resistance for the plateau with integer n was subsequently shown by the mathematical physicist Barry Simon to be equal to a topological invariant. This integer equals the number of holes in the shape representing the quantum Hall state of all the electrons. The connection to topology explained why the Hall resistance was quantized and independent of so many details. It is just like the fact that to a topologist a coffee mug and a doughnut are the same. It also followed that each of the quantum Hall plateaus corresponds to a new state of matter, qualitatively different from other states, just as balls, doughnuts, and pretzels are different from one another topologically. In 1985 von Klitzing was awarded the Nobel Prize in Physics for his discovery of these new states of matter.

Fractional quantum Hall effect

Only two years after von Klitzing's discovery came another surprise. The availability of the combination of higher magnetic fields, temperatures less than one Kelvin, and ultraclean semiconductor heterostructures made it possible to explore a new frontier. With reference to Figure 35 this is the $n = 1$ regime that lies to the right of the curve shown there. Then all of the electrons had the lowest possible energy allowed by quantum theory. Horst Störmer and Daniel Tsui discovered new plateaus in the Hall resistance with values that were not integers but fractions: 1/3, 2/3, 4/3, 5/3,...1/5, 2/5...and so on, with all the denominators of the fractions being odd integers.

These fractions did not make any sense in terms of the picture used to explain the integer quantum Hall effect. You can't have a ball with one-third of a hole in it, or an electron with less than one unit of charge. The explanation of the integer effect assumed that the interactions between electrons due to repulsion of like electrical charges were negligible. But this assumption is debatable when the electrons are in the lowest Landau level as they are closer to one another, and so interact more strongly.

Robert Laughlin developed a new theory that took the interactions between electrons into account and showed that they formed a new quantum state of matter that was only stable for conditions corresponding to the observed fractional values. In his picture, physically it was as if each electron dragged along an even number of quanta of magnetic flux as it moved. This helped explain why the denominators in the observed fractions were always an odd number (3, 5, 7,...). Laughlin's most stunning finding was that if a single electron was added to the 1/3 state the system acted like there were three independent particles, each with a charge of one-third of that of a single electron. It is as if the electron falls apart into three pieces. The properties of the whole system are qualitatively different from those of the individual electrons of which the system is composed. This is an example of emergence. It is impressive that Laughlin provided a theoretical explanation within one year of the discovery of the fractional effect. In contrast, the puzzle of what caused superconductivity was not solved for 46 years. In 1998 the Nobel Prize in Physics was awarded to Störmer, Tsui, and Laughlin.

Quantum spin chains

States of matter that are quantum and topological also exist in crystals composed of chains of magnetic atoms. Unlike most new states of matter, one of these states was predicted theoretically before it was discovered in the laboratory. As we noted earlier, quantum theory describes the magnetic properties of single atoms in terms of a quantity known as spin. Each spin has a direction and its magnitude is quantized; its value can only be an integer or half an integer. A single electron has spin of magnitude 1/2.

The spin of an atom is a sum of the spins of all of the electrons in the atom. In a crystal, the spin of each atom interacts with the spin of the nearby atoms. In ferromagnets, such as iron, this interaction favours parallel alignment of the neighbouring spins. In antiferromagnets, such as nickel oxide, the interaction favours antiparallel alignment. By the 1960s, the nature of these magnetic states in three-dimensional crystals was well understood. At low enough temperatures magnetic order always occurred and properties were the same regardless of whether the quantum spin was an integer or half-integer. This order meant that the directions of all the spins were correlated with one another, regardless of their spatial separation.

In the early 1980s came a big surprise. Chains of magnetic atoms have properties that are qualitatively different from those of a three-dimensional crystal composed of magnetic atoms. Studying theoretical models for chains of atoms where the interaction between neighbouring atoms favoured antiferromagnetism, Duncan Haldane showed that there was a fundamental difference between integer and half-integer spin. For integer spin, the relative direction of the spins could only be correlated over short distances, even at zero temperature. In contrast, in a chain of half-integer spins the direction of the spins is correlated over very long distances.

Haldane's analysis was based on theoretical models that were originally proposed to describe elementary particles known as mesons in nuclear physics. In these models there is a mathematical term that is a topological invariant associated with the abstract mathematical space describing the direction of the quantum spins. Haldane showed that the key difference between integer and half-integer spin chains was whether this topological term has a value different from zero.

Haldane's findings were controversial. However, they were confirmed in the 1990s by calculations on supercomputers and by experiments on crystals composed of chains of magnetic atoms with spin one. It was also shown that although there was no magnetic order in the Landau sense, or spontaneous symmetry breaking, there was an order of topological character. This was manifested in the properties of a finite chain that behaved as if spin-1/2 atoms were at each of the ends of the chain. This split of a spin one into two halves is reminiscent of the quantum Hall effect, where the system acted like electrons split into three parts. In 2010 it was shown that the topological order in both these systems is associated with quantum entanglement, the weird property at the heart of Schrödinger's cat.

Topological insulators

Haldane's creativity continued. In 1988 he published a very abstract paper that showed how the integer quantum Hall effect could occur in the absence of a magnetic field. He studied an artificial model for graphene (which had not yet been made in a lab) in which the electrons moved between atoms differently, depending on whether they were moving forwards or backwards in time. This was related to a problem in elementary particle physics known as the 'chiral anomaly'. Pions are elementary particles that can have a negative, positive, or zero electric charge. The problem is that pions without charge decay almost a billion times faster than charged pions.

Haldane's work achieved little attention until the fabrication of graphene in a lab in 2004 (remember the pencil and sticky tape). This led Charles Kane and Eugene Mele in 2005 to consider a more physically realistic version of Haldane's model and to propose a spin quantum Hall effect in graphene. It turned out that the parameter values they used in their calculations were overly optimistic for graphene. Nevertheless, their work led to other proposals for a topological insulator: a material that is an electrical insulator, but at its surface conducts electricity like a metal. This new state of matter was subsequently observed in several different materials.

Both of Haldane's contributions show the ongoing fruitful interaction between condensed matter physics and elementary particle physics. That interaction began in the 1950s and was central to the BCS theory of superconductivity. In 2016 Haldane shared the Nobel Prize in Physics with David Thouless and Michael Kosterlitz 'for theoretical discoveries of topological phase transitions and topological phases of matter'.

The discoveries of the new states of matter described in this chapter show that condensed matter physics is full of surprises. Like many beloved and successful paradigms, Landau's characterization of states of matter in terms of symmetry breaking and local order turned out to be incomplete. A topological perspective considers the global properties of a system, not just its local properties. This chapter also illustrates how sometimes in science abstract ideas and mathematics such as topology can be relevant to practical issues such as how to define the standard of electrical resistance.

Topological states of matter are examples of emergence. A system of many interacting parts has properties that are qualitatively different from those of the constituent parts. The next chapter will clarify what emergence is and why it matters so much, not just in condensed matter physics, but in many fields of science.

Chapter 9
Emergence: more is different

F. Scott Fitzgerald supposedly remarked to Ernest Hemingway, 'The rich are different from you and me,' prompting the response, 'Yes, they have more money.' This interaction between two of America's famous novelists may be apocryphal, but it embodies the idea that 'quantitative differences become qualitative differences'. This is also a central tenet of Marxist-Leninist ideology, as we pointed out earlier. Similarly, in condensed matter physics qualitative differences, such as different states of matter, arise from quantitative differences such as large numbers of interacting atoms. More is different.

Condensed matter physics is all about emergence: the whole is greater than the sum of the parts. The central question is: how do the physical properties of a state of matter *emerge* from the properties of the atoms of which the material is composed? Earlier chapters discussed emergence with reference to superconductivity, states of matter in Flatland, universality near the critical point in phase diagrams, macroscopic quantum states, and metrology. This chapter focuses on the nature of emergence, its centrality to condensed matter physics, its relevance to other fields of science, and the philosophical questions that it raises.

Characteristics of emergent phenomena

Discussions of emergence in condensed matter physics begin with a seminal article, 'More is Different', published in 1972 by Phil Anderson. He concluded his article with the Fitzgerald–Hemingway exchange and the Marxist idea mentioned above. The concept of spontaneous symmetry breaking was central to his discussion of the hierarchical nature of reality and the limitations of reductionism. The word 'emergence' does not appear in the article, but Anderson later used the term, following discussions with biologists who drew on earlier work in philosophy.

In the social sciences the origin of the concept of emergence is usually attributed to the work of Michael Polanyi in the 1960s. Polanyi began his academic career as a physical chemist, using quantum theory to understand how chemical reactions proceed. After fleeing Nazi Germany in 1933, his interests gradually moved to economics, and then the philosophy of science. Emergence is the title of one of three chapters in his small but influential book *The Tacit Dimension*, published in 1966 and based on lectures that Polanyi gave in 1962 at Yale University.

The characteristics of emergence can be illustrated with an example given by Polanyi of how literature emerges from language (Figure 37). At each level there are distinct components, rules of interaction, phenomena, and concepts. The boxes represent different components of a system. The rules for interactions between these components are given next to the arrows. For example, grammar provides the rules about how words can interact to produce sentences.

An emergent property of a system composed of interacting parts has the following characteristics.

1. An emergent property is not present in the parts.

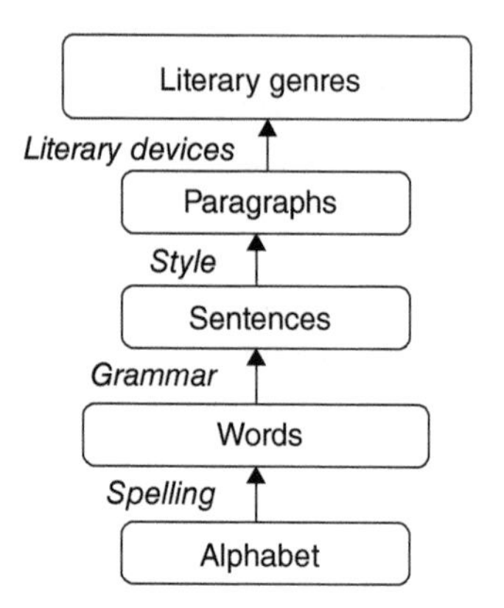

37. Emergence and the hierarchy of levels associated with language and literature.

2. An emergent property is difficult to predict solely from knowledge of properties of the parts and how they interact with one another.
3. An emergent property is associated with a modification of the properties of and the relationships between the parts.
4. An emergent property is universal. It is independent of many of the details of the parts and their interactions.

There are differing interpretations of the nature of emergence, particularly regarding the second characteristic. 'Difficult to predict' is sometimes replaced with 'impossible', 'almost impossible', or 'extremely difficult', or 'possible in principle, but impossible in practice'. After an emergent property has been observed, sometimes it can be understood in terms of the properties of the parts. A key word in the second characteristic above is 'solely'. For example, the BCS theory of superconductivity provided *a posteriori*, rather than *a priori*, understanding of the phenomena.

Examples of properties of a system that are *not* emergent are volume, mass, charge, and number of atoms. These are additive properties. Each property is simply the sum of the properties of the parts.

Emergence in condensed matter physics

In Chapter 1, I stated that emergence is central to condensed matter physics, and referred to irreducibility, surprises, and universality, corresponding to characteristics 1, 2, and 4.

I now illustrate how these four characteristics of emergent properties are related to superconductivity.

1. At low temperatures solid tin exhibits the property of superconductivity. However, a single atom of tin is not a superconductor.
2. Even though the underlying laws describing the interactions between electrons in a crystal have been known for almost 100 years, the discovery of superconductivity in many specific materials was not predicted.
3. In the superconducting state the electrons become ordered in a particular way. The motion of the electrons relative to one another is not independent but correlated.
4. Properties of superconductivity such as zero electrical resistance, the expulsion of magnetic fields, quantization of magnetic flux, and the Josephson effects are universal. The existence and description of these properties is independent of the chemical and structural details of the material in which the superconductivity is observed.

Quasiparticles

An example of emergent entities in condensed matter physics are *quasiparticles*. The concept can be described with the following analogy. When a horse gallops through the desert it stirs up a dust cloud that travels with it. The motion of the horse cannot be separated from the accompanying dust cloud. They act as one entity. Similarly, in a system consisting of many interacting particles, when one particle moves it carries with it a 'cloud' of other particles. This composite entity is referred to as a quasiparticle. It turns out to be easiest to understand the whole system of particles in terms of the quasiparticles rather than in terms of the individual particles.

Quasiparticles are composite objects. Like the constituent particles in the system, quasiparticles each have properties such as charge, mass, and spin. However, these properties of a single quasiparticle may be different from those of the individual particles of which it is constituted. An example is holes in semiconductors; the many electrons in a crystal act collectively to produce a hole (the absence of a single electron), a quasiparticle with the opposite charge to that of a single electron. A more striking example is for the fractional quantum Hall states; the charge of the quasiparticles can be a fraction of the charge on a single electron.

Different musical instruments produce distinct sounds because they are made of different materials, and they vibrate in different ways in response to different stimuli. In general, the vibrations of a medium reflect something about the medium itself. Chapter 3 discussed how in a crystal the number of distinct ways that sound can travel through a crystal reflects the symmetry and ordering of the atoms in the crystal.

When the skin on a drum is hit by a drumstick the skin vibrates at particular frequencies. Similarly, a state of matter responds to external stimuli such as light, sound, or heat, by oscillating at particular frequencies. These vibrations travel through the matter as waves. The properties of these waves reflect the particular order present in the state of matter. Here is a specific example. When a neutron with a particular energy and momentum is absorbed by a ferromagnetic crystal the interaction of the magnetism of the neutron with that of the atoms in the crystal produces a collective oscillation of the magnetic state of the crystal in time and space. Known as a spin wave, this oscillation has a particular frequency and wavelength. In quantum theory, waves and particles are equivalent to one another. The energy and momentum of a particle are related to the waves' frequency and wavelength, respectively. Particles equivalent to light waves are known as photons; particulate equivalents of sound waves are known as phonons. And similarly, the particle equivalent of a spin wave is known as a magnon. These collective excitations are quasiparticles. Whereas the particles in a system may interact strongly with one another, the quasiparticles may interact weakly with one another. This makes analysis and understanding of the relevant theories more tractable.

The quasiparticle concept is a powerful theoretical tool in condensed matter physics. It is the basis for the construction of models that enable emergent phenomena to be understood in terms of the effective interactions between components such as quasiparticles, rather than in terms of the actual constituent particles and their interactions. This approach requires profound physical insight in order to discern what the truly essential components of a system are. Lev Landau was one of the first theoretical physicists to take this approach, introducing the idea of quasiparticles in his theories of superfluidity in ^{4}He and of liquid ^{3}He. This approach was also central to the BCS theory of superconductivity. Phil Anderson was also a master of the approach, using intuition to propose models that were simple enough for analysis and yet complex enough to capture the essential physics associated with a particular state of matter. In 1977 he was awarded the Nobel Prize for work using this approach to understand two specific systems: magnetic atoms in metals and the motion of electrons in materials that are not crystals and are dirty in the sense of containing many impurities.

The scientific challenge of emergence

Over the last century, the scientific strategy known as reductionism has been very successful in providing both a quantitative description and conceptual understanding of a wide range of natural phenomena, by breaking a system down into its parts and studying these. But reductionism is only part of the story. Knowing the constituents of a system of scientific interest and the laws that describe the interactions between the constituents does not mean we can understand or predict the collective properties of the system. Even if we know the exact chemical composition of a crystal and the geometrical arrangement of the atoms in the crystal, it is extremely difficult to predict whether the material will be magnetic, a metal, an insulator, or a superconductor. Indeed, the need to understand emergent properties is central to the biggest challenges in the natural and social sciences today. Knowing the sequence of amino acids that make up a specific protein molecule does not mean we can predict the biological function of that molecule. Knowing the properties of neurons in the brain and their interactions does not mean we can understand consciousness or what thoughts you are going to have. And knowing the psychological profiles of the individual members of a crowd does not mean we can predict whether it will become a violent mob.

The process of reducing complex systems to their component parts and interactions leads to a hierarchical view of reality and scientific disciplines in which physics appears at the base, leading upwards into chemistry, biology, and then to psychology and sociology, increasing in complexity of the system, in terms of number of variables, as we move upwards. At each level there are distinct phenomena, concepts, theories, and scientific methods. Viewing the various disciplines in this way raises questions. Does the fact that physics is on the bottom mean it is the most fundamental discipline? And if we fully understand things on one level, can we actually explain everything on the next higher level? For example, can all of biology really be explained solely in terms of chemistry?

Emergence matters

An emergent perspective on science provides a framework and approach to address scientific questions. It is not just a philosophical issue, but is of practical utility, being relevant to scientific strategy and setting funding priorities. Emergence describes why condensed matter physics works as a unified discipline. As a result of universality there are concepts and theories that can describe diverse phenomena in a wide range of materials.

Emergence makes theory tractable. It is not necessary to keep track of the zillions of atoms in a material to understand its properties. Understanding many aspects of a magnetic phase transition does not require a model that incorporates the quantum theory of all the electrons in the atoms that the material is composed of. Rather, a 'stripped down' model such as the Ising model may be sufficient. In his Nobel Lecture in 2016, Duncan Haldane (who was a Ph.D. student of Phil Anderson) said, 'Looking back,... I am struck by how important the use of stripped down "toy models" has been in discovering new physics.'

Emergence highlights that in a system, whether a crystal, a biological cell, or a community, there are many different scales: size, length, time, energy, and complexity. Looking at the system at these different scales provides complementary descriptions of the system. To investigate a problem, the key is to decide at what scale the system should be studied, and to use appropriate methods for that scale. This strategy has implications for setting priorities, whether for an individual scientist (choice of topic and technique) or a funding agency (goals and budgets for different areas). It is best to invest in a portfolio of complementary approaches that look at different scales, from the microscopic to the macroscopic.

The final chapter considers where condensed matter physics may head in the next few decades. Given that emergent phenomena are hard to predict this may be fraught.

Chapter 10
An endless frontier

What might the future of condensed matter physics be? Just like everyone else, scientists are not good at predicting the future. In 1894 Albert Michelson, famous for his measurements of the speed of light, stated: 'it seems probable that most of the grand underlying principles have been firmly established...' Within two decades physics was turned upside down with the discovery of quantum theory and Einstein's theories of relativity.

Brian Pippard was a distinguished condensed matter physicist who made important contributions to our understanding of metals and superconductivity. In 1961 Pippard was invited to speak at a conference marking the opening of a new IBM research laboratory at Yorktown Heights, New York. Pippard painted a pessimistic view of the future of condensed matter physics:

> the era of the great breakthrough is over... when I suggest to senior physicists that the end of physics as we know it is in sight, they tell me, 'That's just what everybody was saying in 1900'.
>
> ... ask yourselves this question:... what is the most recent discovery in physics that still remains in essence a mystery?... the disappearance of liquid helium, superconductivity, and magnetoresistance from the list of major unsolved problems has left this branch of research looking pretty sick from the point of view of

> any young innocent who thinks he's going to break new ground.... but with the new IBM Laboratory, and all those other labs that we represent, plugging along assiduously doing research, ten years is going to see the end of our games as pure physicists...

Pippard was wrong. Ironically, at the same conference the first results were presented of a macroscopic quantum effect, the quantization of magnetic flux in superconducting cylinders. Furthermore, at the same time Pippard's own Ph.D. student, Brian Josephson, made a discovery that surprised and puzzled many, raised fundamental issues, enabled experimental tests of the foundations of quantum theory, and led to new technologies and metrologies. More discoveries followed: superfluidity in liquid ^{3}He, quantum Hall effects, the renormalization group theory of critical phenomena, topological order, quasicrystals, cuprate superconductors, and graphene, to name a few.

In 1996, a science journalist, John Horgan, published 'The End of Science: Facing the Limits of Knowledge in the Twilight of the Scientific Age', claiming that all the great discoveries had been made in all fields of science and all that remained was to work out minor details and apply scientific knowledge to develop new technologies. But, deciding what is a 'great breakthrough' or 'a major unsolved problem' or 'in essence a mystery' or where the boundary is between 'pure' and 'applied' science involves subjective judgements.

There are a range of views about where condensed matter physics fits into all this. Some would argue that the basic intellectual structure of condensed matter physics is in place and is highly unlikely to be overturned. This may be why at Caltech, Stanford, and the University of Tokyo, condensed matter physics research occurs in the Department of Applied Physics, not in the Department of Physics, reflecting a view that it is not fundamental.

The opposite extreme to pessimism is unbridled optimism manifested in hype. In this final chapter I present an optimistic but sober view, considering some examples of outstanding scientific questions. As condensed matter physics is all about emergence, I anticipate surprises and exciting new discoveries. There is an endless frontier to explore.

Grand challenges

Here are some other outstanding problems and opportunities for breakthroughs, besides the 'holy grail' of superconductivity at room temperature.

Glass and the glass transition. Glasses are everywhere. They are not just in windows. They can be plastics. Even water can form a glass when liquid water is rapidly cooled. Glass is a distinct state of matter. Why are glasses solid? How do they form? We really don't have good answers to such basic questions. In contrast to crystals, which have regular spatial ordering of all the atoms, glasses are characterized by disordered spatial arrangements of the atoms. Understanding the relationship of this structural disorder to the unique physical properties of glasses remains a significant challenge, unlike the well-established understanding of the relationship between structure and properties of crystals. In 1995, Phil Anderson stated that the 'deepest and most interesting unsolved problem in solid-state theory is probably the theory of the nature of glass and the glass transition'.

Schrödinger's cat. Is it dead or alive? Will it be possible to construct a quantum computer, that uses the principles of quantum theory to perform tasks that are impossible on the traditional computers? The most promising approaches are based on SQUIDs using Josephson junctions. But they are still a long way from definitive answers to these questions.

Extreme conditions. The range of temperature, pressure, and magnetic fields that can be explored in laboratories continues to expand. In the past this has led to the discoveries of new states of matter, particularly at low temperatures, high pressures, or high magnetic fields. Examples include the superfluid states of liquid ^{3}He, the quantum Hall effects, and the 18 states of solid water.

Non-equilibrium states. The states of matter discussed in this book are all equilibrium states, that is, they do not change with time. Much is unknown about states that can change with time, particularly when a large external force (e.g. a burst of heat) acts on a system. How will it change in time? Related issues are the formation of patterns, turbulent flow in fluids, and flocking of birds.

Materials by design. One aim of computational materials science is to start with the known chemical composition of a material (i.e. which atoms it is made of) and use the mathematical equations of quantum theory, together with the well-established laws describing the interactions between electrons, atoms, and atomic nuclei, to calculate the physical properties of the material. The availability of ever more powerful computers, new algorithms, and new physical insights has led to significant advances. Commercial software is now routinely used by scientists and engineers to calculate the properties of different materials. Nevertheless, for the most interesting systems, such as cuprate superconductors, these methods fail, sometimes spectacularly. They may not even get qualitative details correct, such as whether a material is a metal, insulator, or a superconductor. These failures result from a multitude of debatable assumptions that must be made to simplify the calculations in order for them to be tractable, even on the most powerful supercomputers. The dream of materials by design is that if an engineer wants a material with specific properties, such as room temperature superconductivity and processible into durable wires that can sustain large electrical currents, these requirements would be entered into a computer that would then calculate the chemical composition, synthesis method, structure, and physical properties of the best candidate materials. Proponents claim that machine learning techniques will facilitate such new discoveries. Whether that will happen remains to be seen.

Interdisciplinarity

Condensed matter physics may continue to make significant contributions to other disciplines, not just to physics, but also chemistry, engineering, computer science, and biology. Condensed matter physics can contribute to these fields with people, techniques, models, and concepts. Complex systems are ubiquitous, both in nature and in society. And as I remarked at the start of the book, condensed matter physics has arguably had the greatest success at understanding emergent phenomena quantitatively, because of the relative simplicity of the systems they study compared to those of fields such as economics.

Soft matter. This book has mostly focused on states of matter that are found in 'hard' materials such as crystals. However, there are other classes of 'soft' materials and states of matter that are 'squishy' such as polymers, membranes, foams, colloids, and liquid crystals. Soft matter is ubiquitous in biological materials and is discussed in detail in the *Very Short Introduction* on the subject by Tom McLeish. He identifies several defining characteristics of soft matter: structures at mesoscopic length scales (i.e. intermediate between the microscopic and macroscopic), universality, common experimental techniques, and multi-disciplinarity. This field is providing new insights into the physics of biological systems and of the science and technology of food. Seminal contributions made by Pierre de Gennes were recognized with a Nobel Prize in 1991. He began his career working on superconductivity and went on to develop a unified framework to understand soft matter, introducing ideas from condensed matter physics such as order parameters, scaling, renormalization, and universality.

Marriage of soft and hard matter. The underlying science of how soft and hard materials interact with one another is poorly understood, whether it is how they do (or do not) bond to one another or how electrical currents can pass through an interface between them. Over the past two decades there has been substantial research (and controversy) into whether electrical circuits could be constructed using DNA. Even the basic question of whether a molecule of DNA was a metal, insulator, or superconductor took a decade to resolve. (It is an insulator.) For biomedical applications it is desirable to marry the 'hard' matter of semiconductor-based electronic devices with the 'soft' matter of biological tissues and cells.

Dirty materials. Most real materials are neither chemically pure nor perfect crystals. They contain impurities and structural disorder. That can be a problem for experiments on condensed matter, because one may not be measuring the properties of the system one is actually interested in. On the other hand, imperfection can lead to interesting new physics and useful technology, particularly when imperfections are well characterized. Silicon Valley may be the birthplace of the semiconductor industry, but pure silicon is actually an electrical insulator. Real transistors are not based on pure silicon but rather silicon containing a controlled number of impurity atoms, such as phosphorus and boron. Light amplification in some optical fibres used in telecommunications is enabled by the presence of small amounts of impurities in the form of atoms of rare earth elements.

Large amounts of impurities can produce new states of matter. In 1958, Phil Anderson showed, to the surprise of many, that a metal containing a large amount of disorder could actually be an electrical insulator. A different class of disordered materials, known as 'spin glasses', are non-magnetic metals that contain a large number of impurity atoms that are magnetic. An example is a sample of gold with one per cent of the atoms replaced with iron. As the temperature is lowered, they undergo a transition to a new state of matter, in which the direction of all the spins of the atoms is frozen. Understanding this magnetic state was a challenge and again Anderson made important contributions, particularly in defining the problem and proposing the simplest possible model that might capture the essential details. But what was the order parameter? What symmetry was broken in this new state of matter?

These questions were answered definitively by Giorgio Parisi, using quite abstract mathematics and drawing on techniques from quantum field theory that he had used previously to understand interactions between quarks in elementary particle physics. The underlying models, physical ideas, and mathematical techniques turned out to be relevant and useful for a wide range of complex systems. This led to new insights into problems in neuroscience, the folding of proteins, biological evolution, computer science, and optimization in applied mathematics. The work on spin glasses is thus another significant example of the cross-fertilization of condensed matter physics with other areas of science. Parisi was awarded a Nobel Prize in Physics in 2021.

Social sciences. The Ising model is a 'stripped down' model designed to capture the essential physics associated with magnetic phase transitions. The model has also found applications in other fields such as neuroscience. Similarly, in the social sciences, simple models (known as agent-based models) have been developed that can be simulated easily on computers. They can describe emergent phenomena such as racial segregation, diffusion of innovation, altruistic cooperation, and epidemics.

Technology

Condensed matter physics has played a significant role in the development of technologies such as liquid crystal displays, superconducting magnets in MRI machines, computer memories, and semiconductor electronics. Future breakthroughs may occur in clean energy including electricity transmission (room-temperature superconductors), energy production (solar cells), energy storage (batteries), and energy conversion (thermoelectric refrigerators). The information technology revolution has followed Moore's law, whereby the number of transistors on a computer chip doubles every two years. This has led to exponential increases in computing power. In 2020, Macbook computers used an Apple M1 chip that contains 16 billion transistors in an area of about one square centimetre. The size of a single transistor is about five nanometres, that is, one twenty-thousandth of the width of a single strand of human hair. Transistors are becoming so small that soon they may not work because different physics, including that associated with quantum theory, becomes relevant. The impending end of Moore's law presents new challenges and opportunities.

The potential of such technological advances is often used to argue for funding of basic research on materials, including in condensed matter. But it should be noted that making a new material or getting a new device to work in a million-dollar laboratory is a long way from producing a widely used commercial product. Regardless of whether it is a photovoltaic cell, a battery, or a superconducting wire, many demanding criteria must be met. Any new technology must be better than any existing technology on numerous counts, such as being cheap to manufacture, scalable to mass production, environmentally friendly, durable, reliable, and safe to manufacture and use.

Opportunities and threats

Condensed matter physics involves a multifaceted approach and I anticipate advances on many fronts will continue: chemical synthesis, device fabrication, experimental techniques, theory, computation, intellectual synthesis, connections with other disciplines, and technological applications.

Discoveries of new states of matter have been enabled by technological advances. For example, discovery of the quantum Hall effects required advances in semiconductor device fabrication, high magnetic fields, and low-temperature refrigeration. Thus, there is synergy with developments in many other fields.

Will there be big new discoveries in condensed matter physics? There are two aspects to this question. First, are there big things to be discovered? If yes, will they actually be discovered?

I believe the answer to the first question is yes for two reasons. First, the past 100 years have given us a continual stream of discoveries, many of them unexpected. Every time that things get a little boring, pretty soon there is something exciting and new. Second, condensed matter physics is all about emergent phenomena in materials. Emergent phenomena are extremely hard to predict. Due to the combinatorics of chemistry, the list of possible materials to study is endless. However, just because such an endless frontier exists does not mean that it will be explored. Successful explorers require courage, creativity, resources, time, and freedom.

I am concerned that the frontiers of condensed matter may not be explored. Reflecting on the contexts in which the pioneers of condensed matter physics, such as Kamerlingh Onnes, Landau, Kapitsa, Anderson, Wilson, Onsager, de Gennes, and Leggett, worked, some common elements were time, stability, job security, mental space, sufficient funding, and intellectual freedom. Anderson spent almost three decades at Bell Labs in its heyday. Thanks to the monopoly of the Bell company in providing telephone services in the USA, it had a secure and stable income, allowing it to provide substantial financial and institutional support for basic research. These pioneers played a long game. They had the freedom to fail, to choose research topics, and to change directions. They did not follow fashion and were fiercely independent thinkers. They had most of the resources they needed and did not have to worry about or fight for funding. Their daily life was very different from that of a researcher today. Their mental space was not filled with an endless stream of distractions and frustrations such as grant proposals, committees, performance metrics, and emails. Most of their time and mental energy was simply focused on curiosity-driven research.

Today running a research group is like running a small business. The skills and personality required for survival, let alone success, may be quite different from those of scientific pioneers. The creativity and technical prowess of the pioneers sometimes also came with introversion and weak skills or low interest in management and communication. I doubt they would survive today, under the intense pressure to produce short-term 'outputs' (papers) and 'impact' (citations) and 'national benefit' (technological, commercial, security, and social). These pressures can lead to researchers working on 'safe' projects in fashionable areas that they are confident will produce results and attention in the short term.

These pressures can also lead some researchers to oversell their research results and the potential commercial applications. Hype is bad for real scientific progress. It wastes precious time and resources, and obscures problems. Determining that a reported research result or area is actually just hype can take significant time. This is particularly true if one actually tries to reproduce a result and discover all the problems associated with it. Resources are mis-allocated. Researchers, students, and funding agencies flock to hyped fields, rather than working on important but less fashionable topics that have robust foundations. Important but difficult problems, such as glass, are put aside. I hope that I am wrong. But I fear that great discoveries may be missed, or at least delayed.

An endless frontier?

Condensed matter physics is all about emergent phenomena, and as we have noted, by definition, emergent phenomena are hard to predict, even if most (or all) of the details of the system components and their interactions are known. Discoveries are often surprising. Almost all discoveries of new states of matter were made by experimentalists, sometimes by accident. Theorists may have had some inklings and suggestions of what to look for, under what conditions, and in what material. However, that is quite different from there being consensus and expectation. Compare and contrast the case of the experimental discovery of the Higgs boson at CERN in 2012. It really wasn't that surprising and there was a strong consensus among theorists both that it would be there and what specific properties it would have.

Serendipity remains a powerful method of discovery. This is why scientists should be given freedom and resources to explore the frontier. What's next? Who knows? All I am game to predict is that condensed matter physics will continue to be an exciting discipline with many surprises and rich intellectual challenges.

References

Chapter 1: What is condensed matter physics?

A. K. Geim, 'Nobel Lecture: Random Walk to Graphene', *Reviews of Modern Physics*, 83, 851 (2011).

Chapter 2: A multitude of states of matter

Hegel proposed three laws of dialectics, the first being 'the law of the transformation of quantity into quality and vice versa'. Engels discussed this in his unfinished 1883 work, *Dialectics of Nature*.

A. G. Spirkin, *The Great Soviet Encyclopedia*, 3rd edition, s.v. 'Transformation of Quantitative into Qualitative Changes'. This entry illustrates the role this 'law' played in Soviet ideology. Retrieved 5 May 2021 from <https://encyclopedia2.thefreedictionary.com/Transformation+of+Quantitative+Into+Qualitative+Changes>.

Chapter 3: Symmetry matters

J. Kepler, *The Six-Cornered Snowflake: A New Year's Gift* (Philadelphia, Paul Dry Books, 2010). Originally written in 1611.

K. Libbrecht, *Snow Crystals: A Case Study in Spontaneous Structure Formation* (Princeton, Princeton University Press, 2021).

R. Cotterill, *The Cambridge Guide to the Material World* (Cambridge, Cambridge University Press, 1985). The quote about X-ray diffraction is on page 56.

Chapter 4: The order of things

F. Close, *Lucifer's Legacy: The Meaning of Asymmetry* (New York, Dover, 2014), revised edition. Chapter 9 discusses the napkin problem and spontaneous symmetry breaking.

The story of Landau's opposition to vortices is described in A. Abrikosov, 'My Years with Landau', *Physics Today*, January 1973, pages 56–60.

N. D. Mermin, *Boojums All the Way through: Communicating Science in a Prosaic Age* (Cambridge, Cambridge University Press, 1990). Chapter 1 recounts the story of how Mermin introduced the term 'boojum' into the scientific literature.

Chapter 5: Adventures in Flatland

E. Abbott, *Flatland: A Romance of Many Dimensions* (London, Sealy & Company, 1884).

J. Gleick, 'Discoveries Bring a "Woodstock" for Physics', *The New York Times*, 20 March 1987, page 1.

Chapter 6: The critical point

K. Wilson, 'Problems in Physics with Many Scales of Length', *Scientific American*, August 1979.

Chapter 7: Quantum matter

G. Gamow, *Mr Tompkins in Wonderland* (Cambridge, Cambridge University Press, 1939).

D. G. McDonald, 'The Nobel Laureate versus the Graduate Student', *Physics Today*, July 2001, page 46. This article recounts the story of Bardeen's opposition to Josephson.

A. J. Leggett, 'Matchmaking Between Condensed Matter and Quantum Foundations, and Other Stories: My Six Decades in Physics', *Annual Review of Condensed Matter Physics*, 11, 1 (2020).

Chapter 8: Topology matters

F. D. M. Haldane, 'Nobel Lecture: Topological Quantum Matter', *Reviews of Modern Physics*, 89, 040502 (2017).

Chapter 9: Emergence: more is different

This dialogue between Fitzgerald and Hemingway is apocryphal. According to 'Are the rich different?' *Financial Times*, 4 December 2014, the quote attributed to Fitzgerald is based on a line in *The Rich Boy*, his 1926 short story: 'Let me tell you about the very rich. They are different from you and me.'

M. Polanyi, *The Tacit Dimension* (Chicago, University of Chicago Press, 2009).

P. W. Anderson, 'More is Different', *Science*, 177, 393 (1972).

Chapter 10: An endless frontier

The quote of Michelson is commonly mis-attributed to Lord Kelvin. See J. Horgan, *The End of Science: Facing the Limits of Science in the Twilight of the Scientific Age* (New York, Broadway Books, 1996), page 19.

B. Pippard, 'The Cat and the Cream', *Physics Today*, November 1961, page 38. This is a transcript of Pippard's invited talk at the conference banquet for the opening of the IBM laboratory.

P. W. Anderson, 'Through the Glass Lightly', *Science*, 267, 1615 (1995).

K. Chang, 'The Nature of Glass Remains Anything but Clear', *The New York Times*, 29 July 2008.

T. McLeish, *Soft Matter: A Very Short Introduction* (Oxford, Oxford University Press, 2020).

Further reading

The recommendations below are not exhaustive but are selected based on my familiarity with them and their potential accessibility to a general audience. Some may provide an entry into the primary scientific literature.

A. Zangwill, *Mind over Matter: Philip Anderson and the Physics of the Very Many* (Oxford, Oxford University Press, 2021).

I. Stewart and M. Golubitsky, *Fearful Symmetry: Is God a Geometer?* (London, Penguin, 1993). Two mathematicians discuss symmetries in nature, including in crystals. Includes many beautiful illustrations.

R. Cotterill, *The Material World* (Cambridge, Cambridge University Press, 2008), 2nd revised edition. This has many beautiful pictures and has a materials science perspective.

R. B. Laughlin, *A Different Universe: Reinventing Physics from the Bottom Down* (New York, Basic Books, 2005). A perspective that emphasizes emergence and discusses many examples from condensed matter physics.

V. Daitch and L. Hoddeson, *True Genius: The Life and Science of John Bardeen: The Only Winner of Two Nobel Prizes in Physics* (Washington, DC, Joseph Henry Press, 2002). Bardeen received Nobel prizes for invention of the transistor and the theory of superconductivity. Why do so few people know of this great scientist, in contrast to Einstein and Feynman?

J. D. Martin, *Solid State Insurrection: How the Science of Substance Made American Physics Matter* (Pittsburgh, University of Pittsburgh Press, 2018). Physics in the USA in the political and economic context of the Cold War is considered, focusing on the rise of solid-state physics, and its metamorphosis into condensed matter physics. Some of the book is summarized in J. D. Martin, 'When Condensed Matter Physics became King', *Physics Today*, January 2019, page 30.

Very Short Introductions. Other titles in this series cover some topics relevant to condensed matter physics in more detail and sometimes from a different perspective.

S. J. Blundell, *Superconductivity: A Very Short Introduction* (Oxford, Oxford University Press, 2009).

S. J. Blundell, *Magnetism: A Very Short Introduction* (Oxford, Oxford University Press, 2012).

A. M. Glazer, *Crystallography: A Very Short Introduction* (Oxford, Oxford University Press, 2016).

T. McLeish, *Soft Matter: A Very Short Introduction* (Oxford, Oxford University Press, 2020).

Websites

<http://www.learner.org/series/physics-for-the-21st-century/emergent-behavior-in-quantum-matter/introduction/>. This is intended for high school students and is written by David Pines.

<condensedconcepts.blogspot.com>. My blog contains many posts on topics and issues discussed in this book.

Index

For the benefit of digital users, indexed terms that span two pages (e.g., 52–53) may, on occasion, appear on only one of those pages.

A

B

C

D

E

F

G

H

N

O

P

Q

R

S

T

U

V

W

X

RELATIVITY

A Very Short Introduction

Russell Stannard

100 years ago, Einstein's theory of relativity shattered the world of physics. Our comforting Newtonian ideas of space and time were replaced by bizarre and counterintuitive conclusions: if you move at high speed, time slows down, space squashes up and you get heavier; travel fast enough and you could weigh as much as a jumbo jet, be squashed thinner than a CD without feeling a thing - and live for ever. And that was just the Special Theory. With the General Theory came even stranger ideas of curved space-time, and changed our understanding of gravity and the cosmos. This authoritative and entertaining *Very Short Introduction* makes the theory of relativity accessible and understandable. Using very little mathematics, Russell Stannard explains the important concepts of relativity, from E=mc2 to black holes, and explores the theory's impact on science and on our understanding of the universe.

www.oup.com/vsi

SCIENTIFIC REVOLUTION

A Very Short Introduction

Lawrence M. Principe

In this *Very Short Introduction* Lawrence M. Principe explores the exciting developments in the sciences of the stars (astronomy, astrology, and cosmology), the sciences of earth (geography, geology, hydraulics, pneumatics), the sciences of matter and motion (alchemy, chemistry, kinematics, physics), the sciences of life (medicine, anatomy, biology, zoology), and much more. The story is told from the perspective of the historical characters themselves, emphasizing their background, context, reasoning, and motivations, and dispelling well-worn myths about the history of science.

www.oup.com/vsi

CHAOS

A Very Short Introduction

Leonard Smith

Our growing understanding of Chaos Theory is having fascinating applications in the real world - from technology to global warming, politics, human behaviour, and even gambling on the stock market. Leonard Smith shows that we all have an intuitive understanding of chaotic systems. He uses accessible maths and physics (replacing complex equations with simple examples like pendulums, railway lines, and tossing coins) to explain the theory, and points to numerous examples in philosophy and literature (Edgar Allen Poe, Chang-Tzu, Arthur Conan Doyle) that illuminate the problems. The beauty of fractal patterns and their relation to chaos, as well as the history of chaos, and its uses in the real world and implications for the philosophy of science are all discussed in this *Very Short Introduction*.

> '. . . Chaos . . . will give you the clearest (but not too painful idea) of the maths involved . . . There's a lot packed into this little book, and for such a technical exploration it's surprisingly readable and enjoyable - I really wanted to keep turning the pages. Smith also has some excellent words of wisdom about common misunderstandings of chaos theory . . .'
>
> popularscience.co.uk

SUPERCONDUCTIVITY

A Very Short Introduction

Stephen J. Blundell

Superconductivity is one of the most exciting areas of research in physics today. Outlining the history of its discovery, and the race to understand its many mysterious and counter-intuitive phenomena, this *Very Short Introduction* explains in accessible terms the theories that have been developed, and how they have influenced other areas of science, including the Higgs boson of particle physics and ideas about the early Universe. It is an engaging and informative account of a fascinating scientific detective story, and an intelligible insight into some deep and beautiful ideas of physics.

www.oup.com/vsi

NUCLEAR POWER

A Very Short Introduction

Maxwell Irvine

The term 'nuclear power' causes anxiety in many people and there is confusion concerning the nature and extent of the associated risks. Here, Maxwell Irvine presents a concise introduction to the development of nuclear physics leading up to the emergence of the nuclear power industry. He discusses the nature of nuclear energy and deals with various aspects of public concern, considering the risks of nuclear safety, the cost of its development, and waste disposal. Dispelling some of the widespread confusion about nuclear energy, Irvine considers the relevance of nuclear power, the potential of nuclear fusion, and encourages informed debate about its potential.

www.oup.com/vsi

CRYSTALLOGRAPHY

A Very Short Introduction

A. M. Glazer

Crystals have fascinated us for centuries with their beauty and symmetry. The use of X-ray diffraction, first pioneered in 1912 by father and son William and Lawrence Bragg, enabled us to probe the structure of molecules, leading to an understanding of their atomic arrangements at a fundamental level. The new discipline, called X-ray crystallography, has subsequently evolved into a formidable science that underpins many other scientific areas.

A. M. Glazer shows how the discoveries in crystallography have been applied to the creation of new and important materials, to drugs and pharmaceuticals and to our understanding of genetics, cell biology, proteins, and viruses.

www.oup.com/vsi

NOTHING

A Very Short Introduction

Frank Close

What is 'nothing'? What remains when you take all the matter away? Can empty space - a void - exist? This *Very Short Introduction* explores the science and history of the elusive void: from Aristotle's theories to black holes and quantum particles, and why the latest discoveries about the vacuum tell us extraordinary things about the cosmos. Frank Close tells the story of how scientists have explored the elusive void, and the rich discoveries that they have made there. He takes the reader on a lively and accessible history through ancient ideas and cultural superstitions to the frontiers of current research.

> 'An accessible and entertaining read for layperson and scientist alike.'
>
> **Physics World**

www.oup.com/vsi